PHYSICS RESEARCH AND TECHNOLOGY

ADVANCES IN APPLIED PHYSICS RESEARCH

PHYSICS RESEARCH AND TECHNOLOGY

Additional books in this series can be found on Nova's website under the Series tab.

Additional e-books in this series can be found on Nova's website under the e-book tab.

PHYSICS RESEARCH AND TECHNOLOGY

ADVANCES IN APPLIED PHYSICS RESEARCH

GABRIEL WATERS
EDITOR

nova publishers
New York

NOTICE TO THE READER

Library of Congress Cataloging-in-Publication Data

ISBN: 978-1-63483-226-7

Published by Nova Science Publishers, Inc. † New York

CONTENTS

PREFACE

This new book presents and discusses current research in the field of applied physics. Topics discussed include radiation effects in piezoelectric AlGaN/GaN heterostructures; fluorescence resonance energy transfer between CdTe quantum dots and organic dye; and band offsets in dielectric/InGaZnO4 thin film transistors.

Chapter 1 – The driving force for GaN electronics is advanced military radar and communications systems. Modern radar and satellite communications operate primarily in the microwave radio regions at frequency bands known as X (8-12 GHz), Ku (12-18 GHz), K (18-27 GHz) and Ka (27-40 GHz) bands. The ability to achieve high power efficiency and compact size are critical for electronics intended for ship, aircraft and satellite based applications. AlGaN/GaN and InAlN/GaN High Electron Mobility Transistors (HEMTs) are well suited to these military electronics needs, with device operating frequencies well within RF bands listed above. These frequencies are readily obtainable with nitride HEMTs because of the high electron velocity in the two-dimensional electron gas (2DEG) channels, while the large breakdown voltages and current densities achievable with nitride heterostructures facilitate high power efficiency. The specific applications for nitride HEMTs include air, land and satellite based communications and phased array radar. The strong bonding in binary and ternary nitrides gives them an intrinsically high radiation resistance. The response of GaN to radiation damage is a function of radiation type, dose, dose rate, carrier density, impurity content and dislocation density. The latter can act as sinks for created defects and parameters such as the carrier removal rate due to trapping of carriers into radiation-induced defects is found to depend on the crystal growth method used to grow the GaN layers. There are three common techniques used to grow GaN layers, namely Metal Organic Chemical Vapor Deposition (MOCVD), Hydride Vapor Phase Epitaxy (HVPE) and Epitaxial Lateral Overgrowth (ELOG), each with its own characteristics in terms of defect density and distribution. The growth method has a clear effect on radiation response beyond the carrier type and radiation source. GaN-based devices are more radiation hard than their Si and GaAs counterparts due to the high bond strength in III-nitride materials. In this paper the authors review data on the radiation resistance of GaN-based transistors such as AlGaN/GaN and InAlN/GaN High Electron Mobility Transistors (HEMTs) to different types of ionizing radiation. In this paper, the primary energy levels introduced by different forms of radiation, carrier removal rates and role of existing defects are discussed. The carrier removal rates are a function of initial carrier concentration, dose and dose rate but not of hydrogen concentration in the nitride material grown by Metal Organic Chemical Vapor Deposition. Proton and electron irradiation damage in HEMTs creates positive threshold voltage shifts due to a decrease in the two dimensional electron gas concentration resulting from electron trapping at defect sites, as well as a decrease in carrier mobility and degradation of drain current and transconductance. Neutron irradiation created more extended damage regions and at high doses leads to Fermi level pinning while ^{60}Co γ-ray irradiation leads to much smaller changes in HEMT drain current relative to the other forms of radiation.

Chapter 2 – In recent years, the light-induced energy transfer process between quantum dots - the organic dye has aroused an interest of many researchers, because quantum dots - organic dye composite system is one of the very promising third-generation photosensitizer and fluorescence biological probe, which has an important significance in the treatment of biomedical tumor. Quantum Dots (QDs) is a kind of novel nano-material. Because of its special size and many special optical properties, it has a wide range of applications in chemical detection, biomedical, electronics. The traditional organic fluorescent dyes, such as Copper phthalocyanine (CuPc) and Rhodamine B (RhB), also have wide applications in the high-technology field due to their special properties of optics, electronics and magnetism. This paper is mainly to study the CdTe quantum dots and organic fluorescent dyes composited system by ultrafast time-resolved spectroscopy technique with two-photon excitation, focusing on the fluorescence resonance energy transfer (FRET) of CdTe QDs-CuPc and CdTe QDs-RhB composite system from theory and experiment. In this article, the main contents and conclusions are as follows: (1) The fluorescence resonance energy transfer in CdTe QDs-CuPc is investigated by ultrafast time-resolved spectroscopy technique equipped with femtosecond laser (780nm,76MHz,130fs). The results show that the fluorescence lifetime of CdTe quantum dots decreases with the increase of CuPc concentration, and the energy transfer efficiency is found to increase with the increase of CuPc concentration. Moreover, the influence of the laser excitation power on the energy transfer efficiency is also studied. It is found that transfer efficiency decreases as excitation laser power increasing. The physical mechanism is the thermal activation with the high power as well as the excited state transitions induced by two-photon excitation. The energy transfer efficiency can reach to be 43.8% when the laser power is 200 mW via two-photon excitation. It indicates that the CdTe QDs-CuPc composite system has high potential as the third generation of photosensitizers. (2) The fluorescence resonance energy transfer between CdTe quantum dots with different sizes and Rhodamine B (RhB) in aqueous solution is investigated by using the time-resolved fluorescence system under two-photon excitation. With the excitation of 800nm femtosecond laser, result shows that the FRET efficiency of the QDs-RhB system increases with the spectral overlap increase of the CdTe emission spectrum and the Rhodamine B absorption spectrum. Specifically, the fluorescence intensity of QDs decreases and the lifetime of QDs decreases while RhB shows the opposite tendency. By means of the Förster Theory of Energy Transfer, the spectral overlap integral J(λ), Foster radius R0 and the FRET efficiency E are calculated and The FRET characteristics of QD-RhB system is characterized. Theoretical analysis reveals that the physical source is the increase of the sample's Forster radius. Moreover, the relationship of the ratio of acceptor/donor concentration and the FRET efficiency is investigated experimentally. When the ratio of acceptor/donor concentration increases, the lifetime of QDs gets shorter, and the FRET efficiency of the QDs-RhB system becomes higher. The two-photon excited FRET efficiency can reach to 40.1% when the concentration of RhB is 3.0×10^{-5}mol/L. It shows that there is a bright future for these materials in bio-probing, bio-imaging, bio-labelling and other biological and optoelectronic applications.

Chapter 3 – In recent years, amorphous oxide semiconductors (AOS) have attracted much attention for use in flat panel displays and flexible thin film transistors (TFTs) due to their improved electrical and optical performance as compared to amorphous Si, which is the dominant technology used in TFTs today. Several AOS materials have been studied including In-Zn-O, In-Sn-O, Zn-Sn-O, Zn-In-Sn-O, and In-Ga-Zn-O (a-IGZO). Among all AOS materials, a-IGZO has been the most widely investigated due to maintaining a very high electron mobility in the amorphous state (10-50 $cm^2 \cdot V^{-1} \cdot sec^{-1}$), and the wide controllability of carrier concentration through oxygen partial pressure during deposition. Moreover, its ability to be deposited at room temperature allows for use of novel flexible substrates, which raises the possibility of making low-cost electronics on a very wide range of arbitrary surfaces including plastic and paper. Amorphous IGZO TFTs are highly transparent, and can be deposited uniformly over large areas, opening up potential applications as switches in the active-matrix and driver-integrated circuits of low cost flexible displays. However, a major remaining issue with IGZO is a better understanding of the appropriate choice of compatible gate dielectrics for devices fabricated using IGZO channels. There are three main criteria that a gate dielectric must possess to be considered acceptable. First, it must be thermodynamically stable with the semiconductor and not react during processing. Secondly, it must provide a high quality interface with low defect and trap density to ensure high carrier mobility. Finally, it must act as a barrier to both electrons and holes, which is among the most important physical parameters for a given heterojunction system. Because these discontinuities can form a barrier for carrier transport across the interface, the knowledge of these quantities is essential for calculating the transport properties of the interface, or the electrostatic potential in a heterojunction device. It is recognized that it is essential to understand the energy band structure at heterojunction interfaces. While there has been much work to date on IGZO devices with various gate dielectrics, there has been very little work on determining the band offsets on such devices. In fact, in two recent reviews of advances in AOS TFTs, it was found that limited gate dielectrics have been studied and more work is needed for next generation TFT devices. The aim of this work is to systematically investigate various high-k dielectrics on IGZO and determine whether there is a sufficient barrier height at the interface in order to minimize the leakage current. This work is important for choosing the proper gate dielectric in next generation of TFT devices.

In: Advances in Applied Physics Research
Editor: Gabriel Waters
ISBN: 978-1-63483-226-7

Chapter 1

RADIATION EFFECTS IN PIEZOELECTRIC ALGAN/GAN HETEROSTRUCTURES

***S. J. Pearton*[1], *Ya-Shi Hwang*[2] *and F. Ren*[2]**
[1]Department of Materials Science and Engineering,
University of Florida, Gainesville, FL, US
[2]Department of Chemical Engineering,
University of Florida, Gainesville, FL, US

ABSTRACT

The driving force for GaN electronics is advanced military radar and communications systems. Modern radar and satellite communications operate primarily in the microwave radio regions at frequency bands known as X (8-12 GHz), Ku (12-18 GHz), K (18-27 GHz) and Ka (27-40 GHz) bands. The ability to achieve high power efficiency and compact size are critical for electronics intended for ship, aircraft and satellite based applications. AlGaN/GaN and InAlN/GaN High Electron Mobility Transistors (HEMTs) are well suited to these military electronics needs, with device operating frequencies well within RF bands listed above. These frequencies are readily obtainable with nitride HEMTs because of the high electron velocity in the two-dimensional electron gas (2DEG) channels, while the large breakdown voltages and current densities achievable with nitride heterostructures facilitate high power efficiency. The specific applications for nitride HEMTs include air, land and satellite based communications and phased array radar. The strong bonding in binary and ternary nitrides gives them an intrinsically high radiation resistance. The response of GaN to radiation damage is a function of radiation type, dose, dose rate, carrier density, impurity content and dislocation density. The latter can act as sinks for created defects and parameters such as the carrier removal rate due to trapping of carriers into radiation-induced defects is found to depend on the crystal growth method used to grow the GaN layers. There are three common techniques used to grow GaN layers, namely Metal Organic Chemical Vapor Deposition (MOCVD), Hydride Vapor Phase Epitaxy (HVPE) and Epitaxial Lateral Overgrowth (ELOG), each with its own characteristics in terms of defect density and distribution. The growth method has a clear effect on radiation response beyond the carrier type and radiation source. GaN-based devices are more radiation hard than their Si and GaAs counterparts due to the high bond strength in III-nitride materials. In this paper we review data on the radiation resistance of GaN-based transistors such as AlGaN/GaN and InAlN/GaN High Electron Mobility Transistors (HEMTs) to different types of ionizing radiation. In this paper, the primary energy levels introduced by different forms of radiation, carrier removal rates and role of existing defects are discussed. The carrier removal rates are a function of initial carrier concentration, dose and dose rate but not of hydrogen concentration in the nitride material grown by Metal Organic Chemical Vapor Deposition. Proton and electron irradiation damage in HEMTs creates positive threshold voltage shifts due to a decrease in the two dimensional electron gas concentration resulting from electron trapping at defect sites, as well as a decrease in carrier mobility and degradation of drain current and transconductance. Neutron irradiation created more extended damage regions and at high doses leads to Fermi level pinning while ^{60}Co γ-ray irradiation leads to much smaller changes in HEMT drain current relative to the other forms of radiation.

INTRODUCTION

The strong bonding in binary and ternary nitrides gives them an intrinsically high radiation resistance [1-13]. The response of GaN to radiation damage is a function of radiation type, dose, dose rate, carrier density, impurity content and dislocation density [11-18]. The latter can act as sinks for created defects and parameters such as the carrier removal rate due to trapping of carriers into radiation-induced defects is found to depend on the crystal growth method used to grow the GaN layers. There are three common techniques used to grow GaN layers, namely Metal Organic Chemical Vapor Deposition (MOCVD), Hydride Vapor Phase Epitaxy (HVPE) and Epitaxial Lateral Overgrowth (ELOG), each with its own characteristics in terms of defect density and distribution. The growth method has a clear effect on radiation response beyond the carrier type and radiation source [12].

There are three main GaN-based heterostructures used for electronics, namely

(i) AlGaN/GaN, which is the most common nitride-based heterostructure
(ii) AlN/GaN produces the highest 2 DEG, but has issues with stability of the AlN.
(iii) InAlN/GaN, an emerging system with high thermal and chemical stability and lattice matching to GaN at an In mole fraction of 0.17. The InAlN/GaN material system offers an attractive alternative to AlGaN/GaN for high power, high frequency applications. The InAlN alloy can be synthesized lattice matched to GaN with large refractive index contrast and sheet charge density roughly twice that of typical AlGaN/GaN HEMTs. This sheet charge density is due to the more than 4x increase in spontaneous polarization of $In_{0.17}Al_{0.83}N$ / GaN as compared to a traditional Al0.2Ga0.8N / GaN HEMT. Although there is no strain-induced piezoelectric component to the overall polarization charge of the lattice-matched InAlN heterostructures, the spontaneous component dominates and leads to a total polarization charge more than 2x that of $Al_{0.2}Ga_{0.8}N$ / GaN. This enables extremely high sheet charge density even in thin (~10nm) barrier HEMT structures. These properties can be exploited to fabricate InAlN / GaN HEMTs with very high current density, low access resistance, aggressive scaling, and monolithic integration of normally-on and normally-off operation.

Given the applications for GaN HEMTs, they will commonly be subject to fluxes of high energy protons and electrons if used in low earth orbit as well as neutrons or gamma rays if used in radiation hard electronics for nuclear or military electronics. Each of these forms of radiation produces different types of damage. In addition, primary defects may recombine, form complexes with each other, with dopants and with extended defects and at high energies the energy of the primary recoils becomes so high that they produce collision cascades and form heavily disordered regions with a very high defect density in the core. In general, proton and electron irradiation produces simple point defects in the nitride lattice [19-26], while neutron irradiation creates extended defects called Gossick zones which are heavily disordered core regions surrounded by a space charge region with strong band bending [11, 12]. The response to gamma irradiation is quite complicated. Compton electrons induced from γ-radiation create electron-hole pairs, thus changing occupancy of traps. Unlike proton irradiation, some studies claim that these defects produced by gamma irradiation can improve device performance such as increasing drain saturation current [15, 16]. In some cases, low dose γ-irradiations partially relaxed the elastic strain in AlGaN/GaN heterostructures and enhanced the electron mobility around 7-8%. This improvement of the electron mobility increased the drain current, whereas others reported the drain current was reduced about 60% after γ-irradiation at around 7 $x10^4$ rad. These defects in that case must reduce the carrier concentration in the irradiated devices, and there was more degradation with increasing γ-irradiation dose. The defects produced by γ-irradiation may be structure sensitive which would be one reason for the discrepancies between different reports. Also, dose clearly plays a role in the performance of the devices after irradiation. Vintusevich et al. [16], found for doses of 10^5 rad, the Ids in the HEMTs increased but start to deteriorate after higher doses of 10^6 rad.

1. Defects in GaN before Irradiation

Native defects are present in high concentrations in GaN and strongly influence its electrical and optical properties [1]. They are often formed as compensation sources when dopants are introduced, or as a result of nonstoichiometric growth or annealing [24-26]. Additional sources of contamination include the gas precursors for crystal growth. The isolated native defects are Ga or N vacancies, interstitials, and antisites (ie. Ga_N or N_{Ga}) but these may interact or form combinations with impurities or dopants to form more complex defects. Schematics of simple vacancies are shown in Figure 1. Both types of vacancies (V_{Ga} and V_N) in GaN are multiply charged defects with several energy levels in the gap. Some studies indicate that the gallium vacancy is the most common native defect in n-type material, while the nitrogen vacancy is the most common in p -type GaN. Due to their larger formation energies, self-interstitials and antisites are less likely to be present in as-grown material [19-22]. There is a strong tendency to self-compensation in nitrides and thus acceptors such as gallium vacancies are favored in n-type material, with donors such as nitrogen vacancies in p-type material. Undoped GaN is usually n-type due to the presence of common impurities such as silicon or oxygen as well as nitrogen vacancies. Based on energetics, the main defects in p-type GaN are predicted theoretically to be V_N, V_{Ga} and N_I, while in n-type material, the most common defects are predicted to be V_N, Ga_N, Ga_I and V_{Ga}.[12, 19, 22].

The structure of most intrinsic defects in GaN is still not well established with the exception of the Ga interstitial [18, 19], the Ga monovacancy [19] and the N split interstitial [22]. Based on simulations, a model for the N split-interstitial defects was proposed as the stable configuration of the N interstitial [27], and only recently has experimental evidence for this defect been reported (Figure 2) [22].

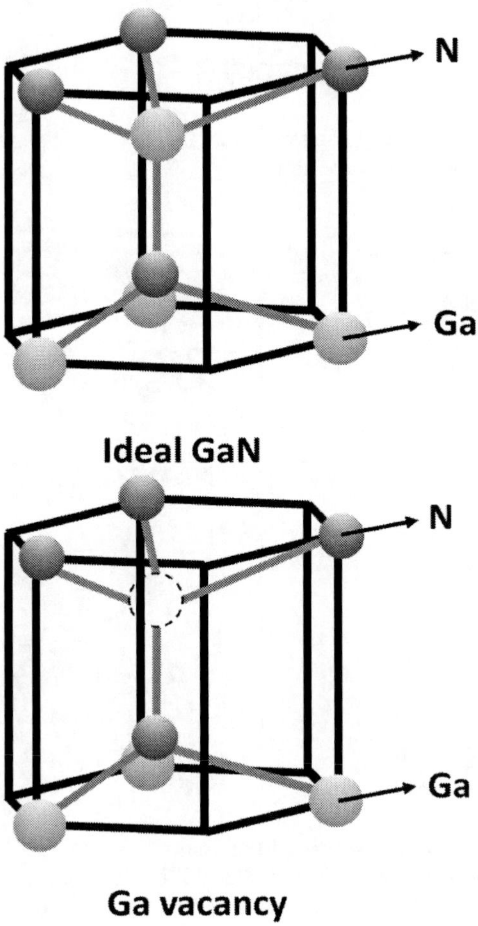

Figure 1. (Continued)

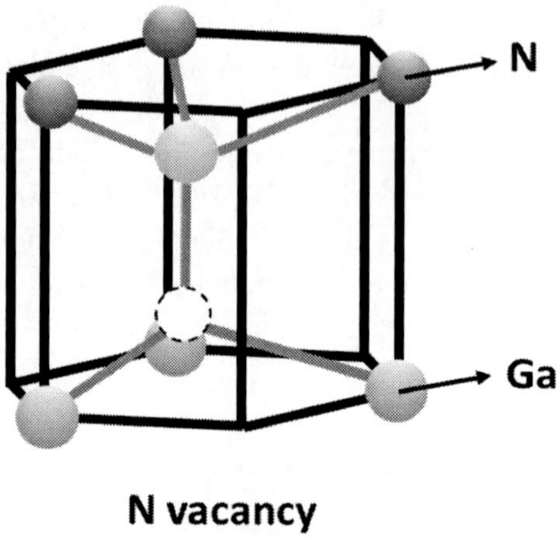

Figure 1. Schematic of the ideal hexagonal GaN lattice (top), a simple Ga vacancy (center) or an isolated N vacancy (bottom).

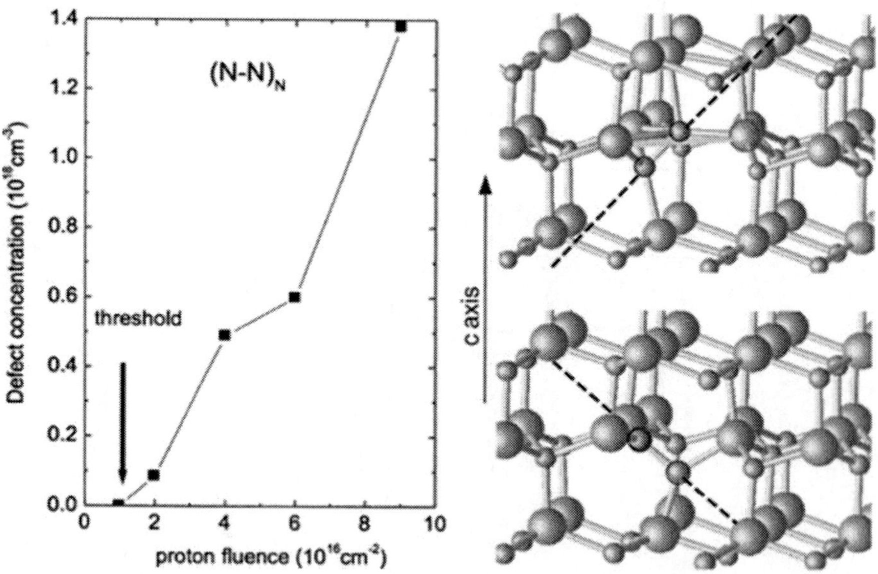

Figure 2. Concentration of the $(N\text{-}N)_N$ split interstitial in GaN as a function of proton dose (left) while at right is shown the ground state (bottom) and excited configurations (top), where one of the N-N dimer atoms has moved through the plane of the 3 Ga ligands. The orientation of the N-N dimers are shown by dashed lines. (after von Bardeleben et al. [22]).

The gallium vacancy is an acceptor with relatively low formation energy in n -type GaN and compensates the n-type conductivity [28-30]. In n -type GaN the Ga vacancy is filled with electrons, and capture of holes during optical injection may lead to radiative transitions of electrons from the conduction band or from a shallow donor to the negatively charged Ga vacancies. These vacancies can easily migrate even at low temperatures and form stable defect complexes [30]. The nitrogen vacancy was long suspected to be responsible for the majority of the native n-type conductivity in GaN to be close to or inside the conduction band, but the concentration of the these vacancies likely depends heavily on growth conditions [1, 2]. The energy level of nitrogen vacancies is reported to be between 0.06-0.13 eV and it is widely accepted that this defect also migrates easily in GaN to form stable defect complexes [1]. However, these probably do not include divacancies such as $(V_{Ga}V_N)$ because of the predicted high formation energy. This defect would be a double acceptor in n -type GaN and a double donor in p-type GaN. Similarly, antisite defects should have a fairly low probability of formation in as-grown crystals because of the large size mismatch between Ga and N atoms [30].

Ga interstitials are also expected to be present only at low concentrations in most as-grown GaN layers because of the high formation energy [27-30], but we should point out that during exposure to radiation fluxes, the formation may be significant. Ga_I is a double donor with an energy level in the gap at around E_C-0.8 eV. Chow et al. [18, 31]. reported the observation of mobile Ga_I in irradiated GaN below room temperature and this high mobility suggests it likely becomes trapped by other defects and is not present as an isolated defect in as-grown layers under equilibrium conditions.

Similarly, the nitrogen interstitial has a high formation energy in as-grown material and may exhibit a number of charge states in the gap [22, 30]. The singly ionized interstitial is an acceptor in n-type GaN with an energy level at around E_C-1.0 eV and will compensate the n-type doping [2, 11]. The migration barrier has been calculated to be low (~1.5eV) and therefore diffusion of the nitrogen interstitials will occur even near room temperature [30]. Recent experimental reports using a combination of electron paramagnetic resonance, density functional theory, and positron annihilation spectroscopy (PAS) have identified the stable nitrogen interstitial defect in irradiated GaN [22]. As expected, the isolated interstitial is unstable and transforms into a split interstitial configuration $(N\text{-}N)_N$. It is generated by particle irradiation with an introduction rate of a primary defect, pinning the Fermi level at the expected value of $E_C-1.0$ eV for high fluences, and anneals out at 400°C [22]. The associated defect, the nitrogen vacancy, was observed by PAS only in the initial stage of irradiatio [22].

Antisite defects on both the Ga and N sub-lattice may also be present. From theoretical considerations, the gallium antisite (Ga_N) is suggested to produce a level at E_V + 0.9 and may be a significant contributor to compensation in p-type material. The nitrogen antisite (N_{Ga}) can behave as a compensating double donor in p -type GaN or an acceptor in n -type GaN but the formation energy is expected to be high [30].

In addition to native point defects, impurities may be present in the as-grown GaN as contamination from precursor gases or the growth reactor itself. The donor impurities include Si, C or Ge when incorporated on Ga sites and O, S, and Se on N sites. The most common of these will obviously be Si and O. It is less common to have acceptor impurities, which could potentially be Be, Mg, Ca, Zn, and Cd on Ga sites or C, Si, and Ge on the N sites [30].

2. Defects in GaN after Irradiation

A simple first–order model for the defects created by ionizing radiation such as protons and electrons involves creation of four levels in the gap, namely, V_{Ga}^- with an acceptor level at E_V+1.0 eV, N_I^- with an acceptor level at E_C-1.0 eV, Ga_I^{++} with a double donor level at E_C-0.8 eV and V_N^+ with a donor level at E_C-0.06 eV [11, 12]. It is important to note that literature reports often cite slightly different energies for these defects, or state them in a different convention. For example, the V_N defect may also be quoted as having an energy level of E_V + 3.28 eV in p-type GaN, which corresponds to the E_C - 0.06 eV value quoted above, although others suggest E_C-0.13 eV as the energy level. Similarly, the V_{Ga} defect may be listed as having an energy level of E_C-2.5 eV, which is equivalent to the E_V+1.0 eV value discussed earlier [11, 12].

Most of the changes in GaN due to proton or electron damage can be understood in terms of introduction of these primary point defects and for example, the carrier removal rate in GaN for light particles is well accounted for by the introduction of these simple defects. At higher doses or when the GaN contains high levels of dopants or other impurities, complexes involving the primary radiation damage defects and these other impurities can occur and the situation becomes more complicated. In addition, differences in electric fields, defect concentrations and experimental precision of the defect spectroscopy techniques can all lead to differences in the experimentally reported values for energy levels in irradiation GaN structures.

Experimentally, a large number of levels are observed to be introduced into both n- and p-type GaN after proton irradiation, as shown in Figure 3.= [23, 26]. These were obtained from a combination of Deep Level Transient Spectroscopy (DLTS) and Deep Level Optical Spectroscopy (DLOS), with the latter allowing measurement in a wider range of the gap. In some cases, the levels are already present in the as-grown material and increase in concentration as a result of the additional defects introduced by the irradiation. Obviously, one would expect that variations in levels observed will occur from different groups because of variations in the starting material and growth technique employed. At this stage there is a still a significant variation in irradiated GaN because of these differences in the quality and purity of material from different sources.

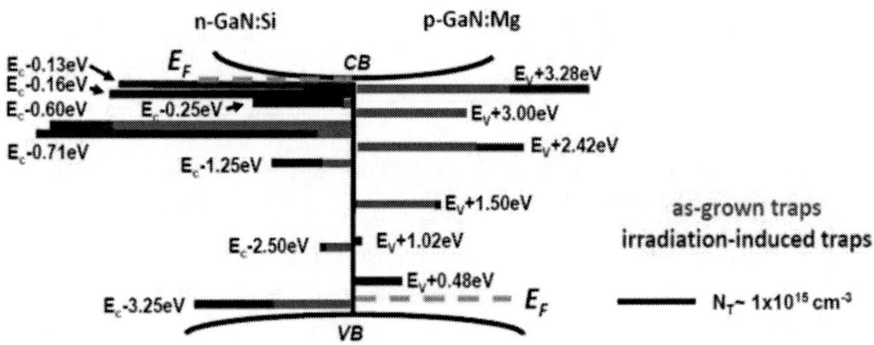

Figure 3. Schematic of position in the bandgap of defects in n- and p-GaN before and after proton irradiation (after Sasikumar et al. [23, 26]).

Similar caveats exist for reporting of carrier removal rates. These are found to be a strong function of initial carrier concentration, dose rate and dislocation density in the material. For example, high dose rates may lead to more extensive defect complexing, or sample self-heating with consequent higher rates of self-annealing of defects. It is important to ensure that carrier removal rates are being measured in a regime in which saturation of the carrier trapping has not already occurred, leading to artificially low extracted carrier removal rates. This will obviously occur at lower doses for lightly doped samples. Figure 4 shows a compilation of removal rates in n-and p-type GaN for different types of radiation as a function of energy.

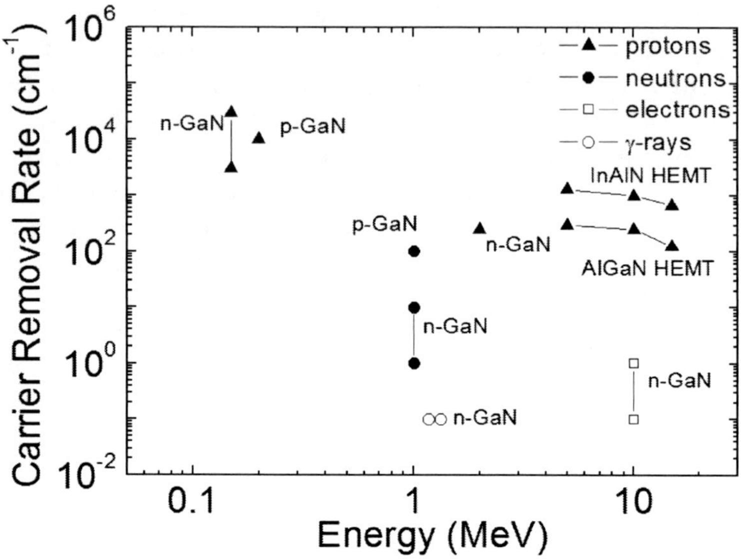

Figure 4. Carrier removal rate in single layer GaN or HEMT structures as a function of energy for different types of radiation.

Proton irradiation leads to carrier removal rate in proton irradiated n-GaN~ 10^2-10^3 cm^{-1} depending on the proton energy and increases for higher donor concentrations [32-41]. In p-GaN implanted with 100 keV protons [2, 11], degradation of luminescent and properties starts at doses ~ 10^{12} cm^{-2}, while decreases in hole concentration were evident for doses beginning at 10^{13} cm^{-2}. Both threshold doses are more than an order of magnitude lower than in proton implanted n-GaN and may involve complex formation between the radiation defects and Mg acceptors. The main deep centers introduced by proton damage in p-GaN have activation energies 0.3 eV, 0.6 eV and 0.9 eV.

3. Changes in HEMT Performance after Irradiation

A wide range of studies have been performed on proton damaged HEMTs in different proton energy regimes [32-46]. For the high energy protons encountered in space-based applications, AlGaN/GaN HEMTs show decreases in tranconductance (gm), drain-source current (I_{DS}), shifts in threshold voltage (V_T) and gate current (I_G) after irradiation with 40MeV protons at doses equivalent to decades in low-earth orbit. These protons create deep electron traps that increase the HEMT channel resistance and decrease carrier mobility. The drain-source currents decreased 15-20% for irradiation with a dose of 5 x 10^9 cm^{-2} and 30-50% decrease at a dose of $5x10^{10}$ cm-^{2}. Post-irradiation annealing at 300°C restored 70% of initial gm and Ids in such HEMTs receiving doses of $5x10^{10}cm^{-2}$. For proton-irradiation at 1.8 MeV, both AlGaN/GaN and AlGaN/AlN/GaN HEMTs showed significant degradation, around 40% drain current reduction, after the devices were exposed to protons at a dose of 10^{14} cm^{-2}. Sasikumar et al. [23, 26], have identified two main levels as being responsible for the threshold voltage shifts in irradiated AlGaN/GaN HEMTs, namely at E_C-0.3.25 eV and at E_C-0.1.25 eV. The latter is possibly the nitrogen interstitial defect, while the former is not firmly identified.

The functional dependence of HEMT degradation on proton energy is interesting because of the fact that the active layers of the device are so close to the surface. For this reason, higher energy protons actually create less displacement damage or non-ionizing energy loss in the active regions and create less degradation of the HEMT electrical performance [47-49]. For example, drain-source current- voltage (I_{DS}- V_{DS}) characteristics at various gate-source voltage (V_{GS}) obtained before and after irradiation at various proton energies (5, 10 and 15 MeV) are shown in Figure 5. The dose was fixed at $5x10^{15}$ cm^{-2}. The change of knee voltage was nominal after proton irradiation. However, a reduction of the drain current level was observed from all of the proton–irradiated AlGaN/GaN HEMTs. As expected from Stopping and Range of Ions in Matter (SRIM) results, there was less damage from higher proton energies since most protons will penetrate the samples including the shallow two dimensional electron gas (2DEG) layer and cause less damage. The drain-source voltage at 6 V and zero gate voltage was reduced by 9% for 15 MeV irradiation, 15% for 10 MeV irradiation and 47% for 5 MeV irradiation.

For 10 MeV electron irradiation of HEMTs, the observed carrier removal rates are lower than for protons of comparable energy. The 50% degradation of the 2 DEG conductivity happens at several times higher doses (close to 3×10^{16} cm^{-2} versus 6.5×10^{15} cm^{-2}) for AlN/GaN than for AlGaN/GaN structures, while an even lower fluence of 1.3×10^{16} cm^{-2} was needed for InAlN/GaN HEMTs, in line with previous observations for neutron irradiated heterojunctions. The shift of C-V characteristics was due to an increased concentration of deep acceptor traps in the barrier/interface region. In AlGaN/GaN/Si transistors there an increase in concentration of deep barrier/interface traps with activation energy of 0.3 eV and 0.55 eV. This increase in trap density correlates with an increase in gate lag of HEMTs after electron irradiation. Electron irradiation at an energy of 10 MeV of n-GaN grown by both MOCVD and ELOG led to compensation of n-type conductivity and carrier removal rate increased substantially with starting donor concentration.

The main compensating defect created was an electron trap at E_C- 0.15 eV. Once the dose is high enough and the Fermi level crossed this level, two centers at 0.2 eV and 1 eV contribute to compensation. After very high doses, the Fermi level in moderately doped GaN is pinned near Ec-1 eV, which again appears to be due to the nitrogen interstitial.

The carrier removal rate in ELOG n-GaN was lower than MOCVD samples with similar doping levels. As with proton irradiation, AlN/GaN heterostructures were more resistant to electron irradiation damage hard than AlGaN/GaN or InAlN/GaN.

For fast neutrons which create large recoil cascades, carrier removal in HEMTs is by disordered regions in which the Fermi level in the core is pinned between the Ga interstitial donor level and the N interstitial acceptor level, ie between E_C-0.8 eV and E_C-1.0 eV, surrounded by a space charge region with a high potential barrier for electrons.

Carrier removal rate is experimentally found to be related to the threading dislocation distribution. The major electron traps created have levels at E_C-0.45 eV, 0.85 eV (Ga_I^{++}), E_C-1 eV(N_I^-). The carrier removal rate is found to be the sum of contributions from these disordered regions and from traps whose introduction rate increases with doping and decreases in the order MOCVD/ELOG/ HVPE GaN.

Recent experiments using Electron Paramagnetic Resonance (EPR) have found strong evidence of a nitrogen sublattice defect produced with high introduction rates in GaN created by proton, electron or ion irradiation at room temperature.

The defect configuration is actually that of a $(N\text{-}N)_N$ split interstitial which is electrically active and pins the Fermi level at E_C-1.0 eV below the conduction band. This N split interstitial is removed by annealing at 400° C. As pointed out above, this defect must be considered along with the N vacancy donors, N interstitial acceptors, and Ga vacancy acceptors [22]. The new experimental data suggest the N split interstitial is ambipolar, i.e., a deep acceptor defect with a level at Ec-1:0 eV in n-type material and a deep donor in p-type material. As its introduction rate is close to the theoretical value for primary N interstitials, the transformation in the split interstitial is apparently complete in n-type and semi insulating [22].

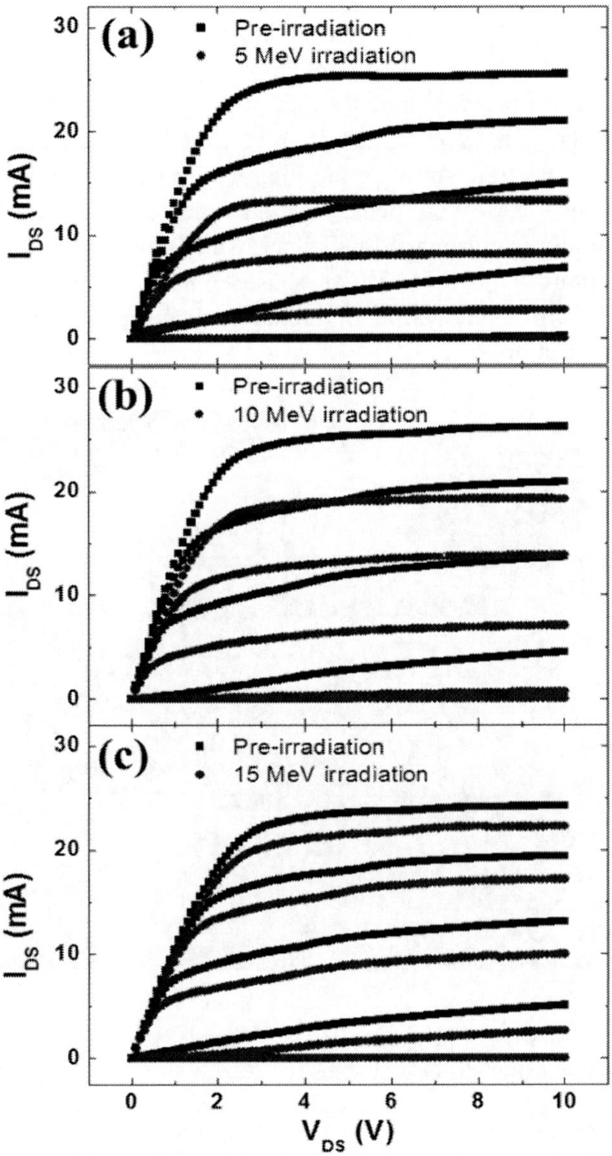

Figure 5. I_{DS}-V_{DS} before and after proton irradiation at energies of (a) 5 MeV (b) 10 MeV (c) 15 MeV. The dose was fixed at $5x10^{15}$ cm^{-2}, while V_{GS} = 0 V ~ -4 V.

4. Role of Growth Technique and Extended Defect Density

At this stage, GaN heterostructures for transistors are usually grown on lattice mismatched substrates and contains a high density of dislocations. In GaN HEMTs grown on sapphire substrates, the dislocation density is ~10^8-10^9 cm^{-2}. Figure 6 shows a TEM cross-section of an AlGaN/GaN HEMT structure grown on a SiC substrate showing the high dislocation density, still typically >10^{7} cm^{-2}. The dislocation density can be decreased to ~ 10^6 cm^{-2} for epitaxially laterally overgrown ELOG GaN, while thick (>200 μm) GaN films grown on sapphire by hydride vapor phase epitaxy HVPE have dislocation densities < 10^7 cm^{-2}. Interaction of radiation defects with dislocations is a factor in all these approaches.

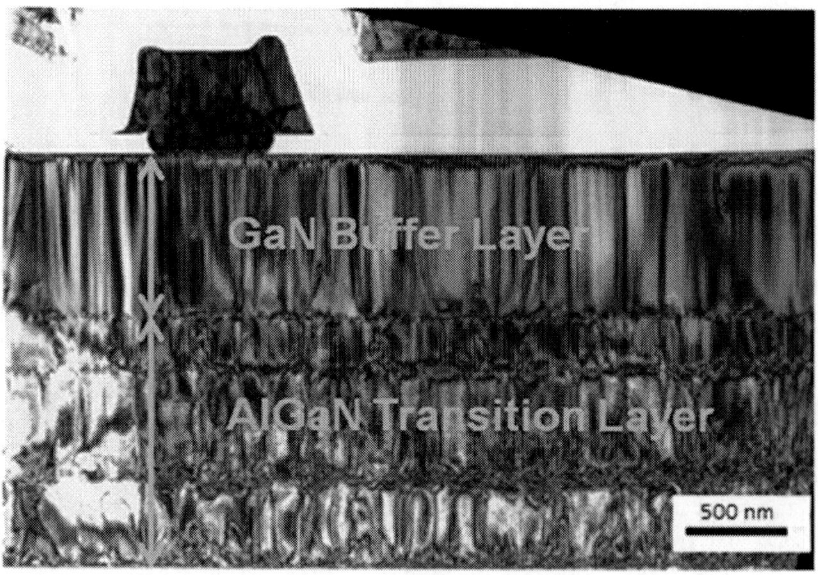

Figure 6. Cross section TEM image of AlGaN/GaN HEMT structure grown on a SiC substrate showing the high dislocation density.

One of these effects is spatial correlation of electron capture by deep traps decorating dislocations [2, 11, 12], including the 1 eV N_I-related acceptors, 0.8 eV Ga_I and 0.95 eV V_{Ga} electron traps introduced by 2 MeV proton irradiation. During irradiation some of the interstitial defects can migrate to dislocation boundaries. Another effect is the impact of dislocations on mobility of charge carriers. With increasing radiation dose, the electron concentration within the grains becomes lower and the potential barriers is higher. For p-GaN, tunneling of holes is difficult even at high acceptor densities and hole mobility seldom shows the expected theoretical temperature or concentration dependence [11]. The theory predicts a strong increase of the electron mobility with decreasing dislocation density and increasing electron concentration, the latter due to enhanced screening of dislocations. However, the effect of dislocations on the carrier mobility is not reduced to the sum of contributions from individual dislocations. For dislocation densities exceeding $\sim 10^8$ cm^{-2} in GaN, dislocations form a cellular structure with dislocation boundaries formed mostly by the threading edge dislocations and characteristic grain size determined by the dislocation density. For carriers to travel over the grain boundaries involves overcoming a high potential barrier. When the electron concentration is high, tunneling through the barriers is efficient and the material behaves as though electrical nonuniformities were not present.

As an example, neutron irradiation of undoped ELOG GaN resulted in a five times lower effective removal rate than for standard MOCVD GaN [11]. The carrier removal rates for thick HVPE grown GaN were lower than for either the MOCVD or ELOG material and the dependence of the removal rate on starting concentration wass stronger. This is not a consequence of dislocation density changes, because the dislocation density in the low-dislocation-density wing of ELOG samples is similar to the dislocation density in bulk HVPE samples but appears to be related to the uniformity of dislocation distribution. For MOCVD GaN the dislocations form a clear cellular pattern, with the typical size of dislocation cell of 0.3-0.5 μm. In ELOG GaN, wing regions with randomly distributed dislocations with density $\sim 5 \times 10^6$ cm^{-2} are adjacent to the high dislocation-density window regions, whereas in HVPE samples the dislocation density is uniformly low at the level of$\sim 10^7$ cm-^{2}. This could be important for the probability of primary radiation defects to escape to the dislocations. It is well known that Ga vacancies (V_{Ga}) and Ga interstitials (Ga_I) are mobile at room temperature in GaN. We also note that it has been discovered by electron beam induced current (EBIC) profiling in neutron irradiated ELOG n-GaN that the decrease of the charge collection efficiency in EBIC occurs much more rapidly in the parts of ELOG wing region close to the window region than in the bulk of the ELOG wing where the radiation induced changes are the lowest [11, 12]. Carrier removal rates in neutron irradiated n-GaN increase with doping and decrease in the sequence MOCVD/ELOG/HVPE. The former were associated with disordered regions in GaN and determine the carrier removal rate in undoped films. The latter were attributed to radiation defect complexes with shallow donors.

Figure 7 shows the relative decrease of electron mobility as a function of proton dose for AlGaN/GaN HEMTs grown on different substrates (from Koehler et al. [25]). Note that mobility is degraded more for GaN HEMTs on SiC than sapphire. The higher the initial defect level, the more insensitive it is to radiation. One final issue with effect of growth method is the presence of hydrogen in material grown by MBE using NH_3 plasmas as the nitrogen precursor. That is covered in the next section.

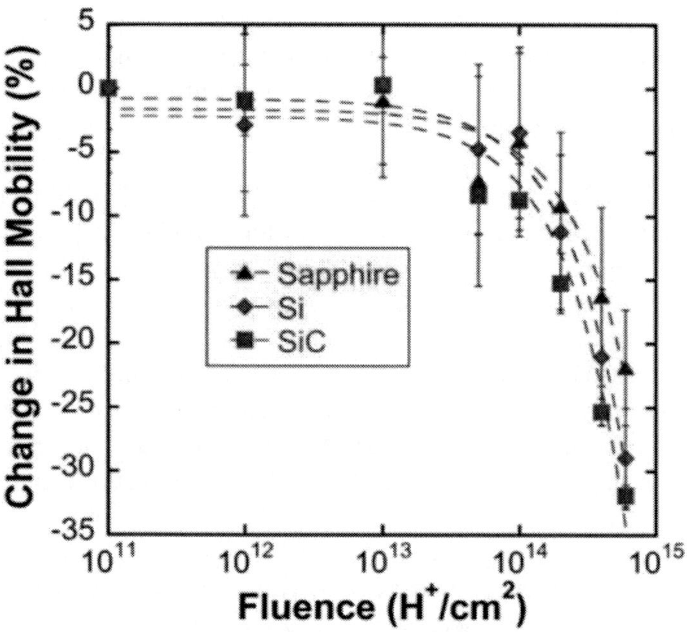

Figure 7. Degradation of electron mobility as a function of proton dose for AlGaN/GaN HEMTs grown on different substrates (after Koehler et al. [25]).

5. Effects of Hydrogen and Electric Field on Damage Creation in HEMTs

Ionizing radiation slowly degrades the properties of oxide-semiconductor structures in Si. Electrons and holes released by radiation interact with defect sites and hydrogen species in oxides, trapping charge and releasing protons, creating permanent oxide-trapped charge and charged interface traps. This leads to undesirable effects in the electrical characteristics of devices, eg. irradiated MOS devices exhibit negative shifts in threshold voltage and large increases in leakage current, and bipolar transistors show increased base current [50]. Similar behavior might be expected in dielectric layers that passivate the surface of GaN-based heterostructures, but little work has been performed to date.

The mechanisms involved in oxide radiation damage have been the subject of numerous studies [49], but they are agreed to be some variant of the two-stage hydrogen model [50]. In Si-based MOS devices, the ionizing radiation releases holes and electrons with the holes becoming trapped at defects, forming the well-known E' centers in oxide. Some of these centers remains as permanent oxide-trapped charges. The interstitial molecular hydrogen cracks on charged vacancies, creating hydrogenated vacancies and protons, the latter which migrates to the interface to depassivate Si-H bonds at the Si/SiO_2 interface, creating interface traps. The vacancy complexes also act as recombination centers, believed to be responsible for effects such as Enhanced Low Dose Rate Radiation Sensitivity (ELDERS). The presence of hydrogen at MOS gate interfaces has a significant influence on interfacial trap formation and density in Si-based devices. The most common form of introducing hydrogen into GaN devices occurs during steps like implant isolation. For hydrogen doses into GaN at doses exceeding $\sim 10^{16}$ cm^{-2}, formation of hydrogen bubbles becomes an issue.

Most GaN-based HEMT structures still employ metal gates, not MOS gates and therefore hydrogen-related effects at the gate-semiconductor interface are minimized. Since the MOS gate approach affords many advantages, including better thermal stability, higher voltage swings and potentially improved reliability, the topic of radiation effects in MOS-gate HEMTs will grow in importance. The lower on-resistances of GaN-based HEMTs relative to Si transistors are due to the high carrier mobility in the two-dimensional electron gas (2DEG) channels and also the large critical electric fields. High breakdown voltage AlGaN/GaN devices have been demonstrated and have shown on-resistances below those achieved in Si. A typical breakdown voltage for HEMTs with a common layer thickness of 2 um is roughly 800V. To increase this voltage requires methods such as acceptor doping of the buffer layer, increasing the active layer thickness, use of a recessed gate or increasing the bandgap of the buffer layers through use of AlGaN. Another approach to increasing gate breakdown voltage is to use a metal-insulator semiconductor HEMT (MISHEMT) structure rather than the conventional Schottky gate. While Schottky metal gate HEMTs offer the highest frequency performance, issues such as current collapse, high gate leakage, and reliability problems related to the stability of the metal gate have limited the potential for high power applications. Many different dielectric layers have been demonstrated for GaN-based metal-oxide-semiconductor (MOS) or MISHEMTs, including SiO_2, AlN, SiN_x, Al_2O_3, Sc_2O_3, HfO_2, La_2O_3, and TaO_xN_y. Use of a gate dielectric with high permittivity produces more efficient gate modulation and thus a smaller decrease in transconductance and threshold voltage shifts.

We have found that hydrogen incorporated into the active regions of HEMTs grown by MOCVD does not change their relative response to proton damage. AlGaN/GaN and InAlN/GaN HEMTs were exposed to deuterated NH_3 plasmas to load deuterium and hydrogen into the near surface region, as shown in the SIMS profiles in Figure 8. There was no significant change in the amount of degradation in drain current for devices that were hydrogenated by exposure to the ammonia plasmas. However, in the case of MBE grown GaN HEMTs, simulations of hot-electron degradation in these devices shows that under semi-ON state bias conditions, the average electron energy is sufficiently high that when transferred to the lattice can cause defect dehydrogenation and device degradation. The responses to 1.8 MeV proton irradiation of AlGaN/GaN HEMTs grown by MBE under Ga-rich, N-rich, and NH_3-rich conditions showed that the NH3-rich devices were more susceptible to proton irradiation than the Ga-rich and N-rich devices. The changes in defect concentrations were consistent with the reconfiguration and/or dehydrogenation of oxygen-related defects in Ga-rich devices. In these types of HEMTs, the generation of active defects is increased by the release of hydrogen atoms that passivate pre-existing defects. Density functional theory calculations showed that N vacancies and Ga-N divacancies lead to enhanced noise in these devices.

Only limited studies have been carried out on the effect of electric field on the amount of damage created in proton irradiated HEMTs. Over the usual range of gate and drain-source operating voltages, there was no significant effect of bias during irradiation on the amount of device degradation in AlGaN/GaN HEMTs and there was also no effect of field plates which reduce the peak electric field near gate edges.

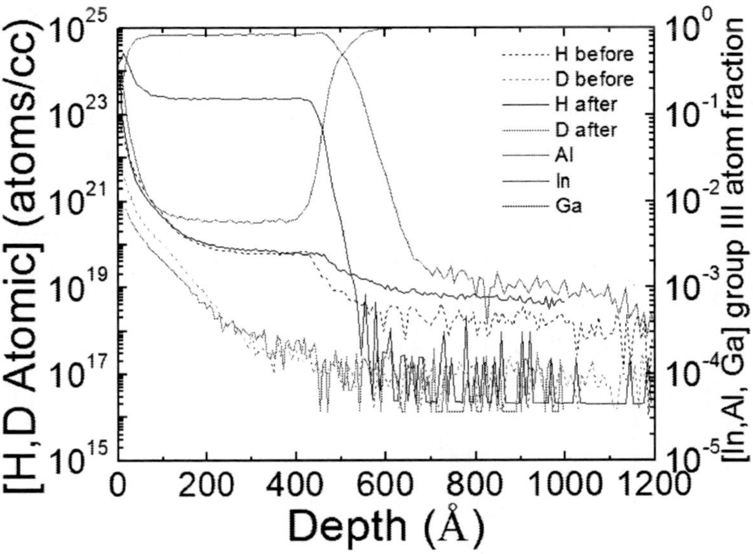

Figure 8. Hydrogen and deuterium profiles in InAlN/GaN HEMTs exposed to isotopically enriched NH_3 plasmas prior to irradiation with 150 keV protons to a dose of $5x10^{14}$ cm^{-2}.

6. Use of Proton Damage to Enhance Off-State Breakdown Voltage

Proton irradiation from the backside of the samples has been observed to enhance off-state drain breakdown voltage of AlGaN/GaN HEMTs grown on Si substrates. Via holes were fabricated directly under the active area of the HEMTs by etching through the Si substrate for subsequent backside proton irradiation [36]. By taking the advantage of the steep drop at the end of proton energy loss profile, the defects created by the proton irradiation from the backside of the sample could be precisely placed at specific locations inside the AlGaN/GaN HEMT structure. There were no degradation of drain current nor enhancement of off-state drain voltage breakdown voltage observed for the irradiated AlGaN/GaN HEMTs with the proton energy of 225 or 275 keV, for which the defects created by the proton irradiations were intentionally placed in the GaN buffer. HEMTs with defects placed in the 2 dimensional electron gas (2DEG) channel region and AlGaN barrier using 330 or 340 KeV protons not only showed degradation of both drain current and extrinsic transconductance but also exhibited improvement of the off-state drain breakdown voltage. This is shown in Figure 9 and Table 1. FLOODS TCAD finite-element simulations were performed to confirm the hypothesis of a virtual gate formed around the 2DEG region to reduce the peak electric field around the gate edges and increase the off-state drain breakdown voltage [36].

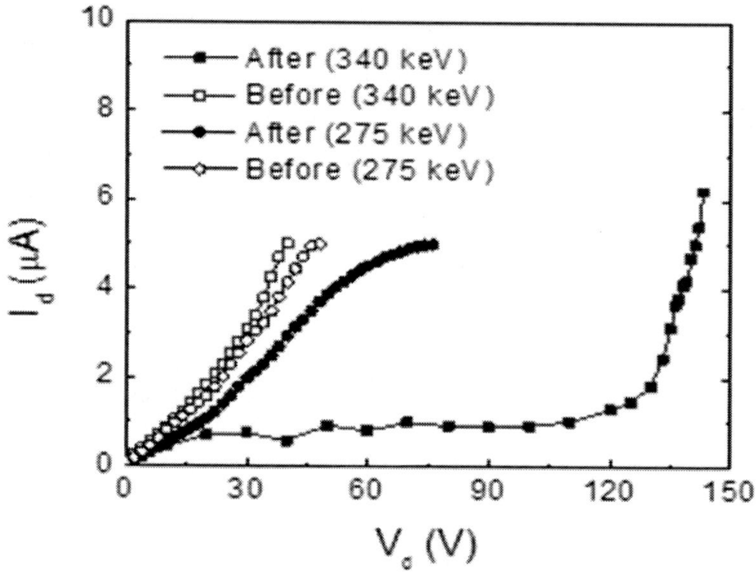

Figure 9. Off-state drain breakdown voltage of AlGaN/GaN HEMTs before and after irradiation with a proton energy of 340 keV and a dose of 5×10^{12} cm^{-2}.

Table 1. Summary of the reduction in I_{DSS} and improvement of off-state drain breakdown voltage, V_B, after proton irradiation with various energies and doses

Proton Dose (cm^{-2})	Energy (keV)	ΔI_{DSS} (%)	ΔV_{RB}
4×10^{12}	225	4	0
4×10^{12}	275	6	0
7×10^{12}	330	-36	151
1×10^{12}	340	-4	9
5×10^{12}	340	-15	136

7. Annealing of Radiation Damaged HEMTs

Annealing of defects introduced in single layers of n-GaN by 2 MeV protons or 0.2-2.4 MeV electrons showed that some of the defects started annealing around 250°C and was complete by 350°C [11, 12]. Electron traps associated with N interstitials annealed out at 380°C. The Ga interstitial-related deep donors started annealing even at room temperature, while the 1 eV Ga vacancy states were stable up to 500°C. When the proton or electron doses are increased or neutrons are used for the irradiation and the density of defects is increased, the thermal stability of radiation damage becomes higher. For example, in n-GaN irradiated with high doses of protons, the bandedge luminescence intensity could not be restored even after annealing at 800°C. Similarly, the evolution of the sheet resistivity of n- GaN irradiated with fast and thermal neutrons to high fluences (~10^{17} cm) showed a number of recovery stages. The as-irradiated resistivity decreased at the 150-250°C stage, increased strongly from 250-450°C and then gradually decreased in a broad stage between 500-1000°C [11]. The first stage corresponds to reconstruction of common deep acceptors which explains the decrease in resistivity. The reverse annealing stage at 250-450°C is most likely due to movement of the Ni, Gai centers forming new deep compensating centers. The onset of the third stage of recovery at 500°C correlates with the V_{Ga} acceptors annealing stage which explains the decrease of the resistivity. The Fermi level was pinned at deep centers with activation energy 0.45 eV. The most prominent electron traps were at 0.9 eV and 1 eV, likely related to the Gai donors and the Ni acceptors, but with a high binding energy, possibly trapped within disordered regions. After 1000°C annealing the Fermi level was pinned near Ec-0.2 eV and DLTS spectra were dominated by the 0.6 eV and 0.9 eV traps in high concentration. The total concentrations of the 0.45 eV traps pinning the Fermi level after 800oC annealing and of the 0.2 eV traps dominant in the 1000°C annealing are close to each other and equal to the number of donor Ge atoms converted from Ga by interaction with thermal neutrons (2×10^{16} cm^{-3}). Hence, these relatively deep traps could be complexes of radiation defects with donor atoms [11, 12]. Even after moderate neutron doses, removal of the disordered regions was incomplete and the initial conductivity was not restored.

For very high neutron fluences the resistivity of GaN passes through a maximum related to the onset of hopping conductivity. The activation energy for the temperature dependence of resistivity for doses before the maximum resistivity showed the usual value of 0.9-1 eV. After a fluence corresponding to maximum resistivity, the temperature dependence was weaker and annealing showed a strong reverse annealing stage up to 300°C where the density of radiation defects decreased and the activation energy returned to the 0.9 eV value. Complete recovery could not be attained even after annealing at 1000°C [11].

For AlGaN/GaN HEMTs irradiated with 40MeV protons at doses equivalent to decades in low-earth orbit, post-irradiation annealing at 300°C restored 70% of initial gm and Ids in HEMTs receiving doses of $5x10^{10}cm^{-2}$. In AlN/GaN HEMTs irradiated with 5 MeV protons at fluences from $2x\ 10^{11}$ to $2x\ 10^{15}$ protons/cm^2, there was less degradation in drain current as a result of the more radiation-hard heterostructure compared to AlGaN/GaN [51]. Annealing of these devices at 250-350 °C from 5 minutes to 3 hours showed that the forward and reverse gate current decreased 10% under these conditions, moving back towards their pre-irradiation values, but there was no obvious drain current improvement obtained. Proton irradiation of Sc_2O_3/GaN and Sc_2O_3/MgO/GaN MOS diodes at two energies, 10 and 40 MeV, and total fluences of $5x10^9\ cm^{-2}$ caused a decrease in forward breakdown voltage and a flat-band voltage shift in the capacitance-voltage characteristics, indicating a change in fixed oxide charge and damage to the dielectric [52]. The interface state densities after irradiation increased from $5.9x10^{11}\ cm^{-2}$ to $1.03x10^{12}\ cm^{-2}$ in Sc_2O_3/GaN diodes and from $2.33x10^{11}$ to $5.3x10^{11}\ cm^{-2}$ in Sc_2O_3/MgO/GaN diodes. Post-annealing at 400°C in forming gas recovered most of the original characteristics but did increase the oxide/GaN interfacial roughness. These results were not dependent on the proton energy, but the radiation did enhance roughening of the interface compared to unirradiated diodes annealed in the same fashion, where no significant change in interface roughness was observed until annealing of 800C.

AlGaN/GaN HEMTs irradiated with ^{60}Co gamma-ray doses up to 600 Mrad of 60Co –rays showed an increase in reverse breakdown voltage (V_{RB}) by a factor of two, a negative shift in V_T and a decrease in g_m by 25%. These results are consistent with the gamma-irradiation causing a decrease in the effective channel doping through introduction of deep electron traps. The irradiated devices were annealed at 200° C for 25 minutes and these HEMTs showed partial recovery of device performance. The impact of irradiation and annealing on minority carrier diffusion length and activation energy were monitored through the use of the Electron Beam Induced Current (EBIC). The level of recovery of gamma irradiated devices after annealing treatment depended on the dose of irradiation [10]. Recovery of devices was evident from the significant increase in the diffusion length after annealing, as shown in Figure 10. Gamma-irradiation induced nitrogen vacancy defects created additional traps in AlGaN/GaN HEMTs, which decrease the carrier concentration and mobility and increase the activation energy related to carrier recombination. These defects act as scattering centers which result in degradation of both device performance and reliability. After annealing the device at 200 ºC, the defects were reduced, resulting in an increase of both diffusion length and performance.

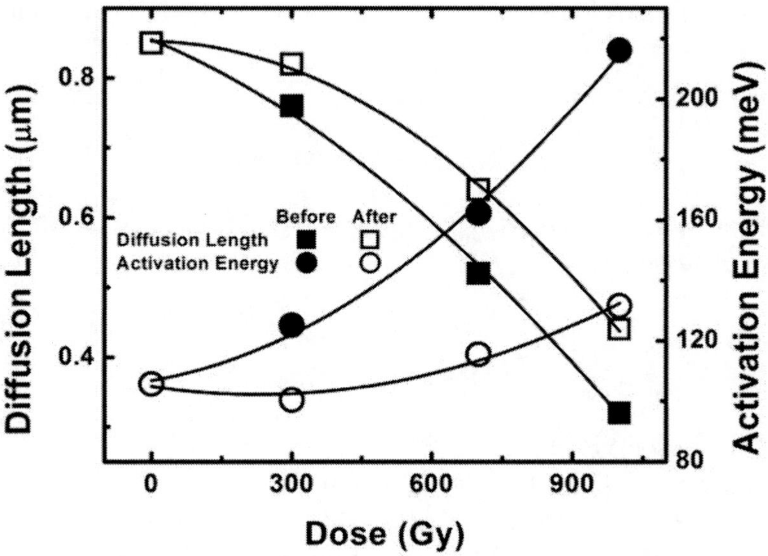

Figure 10. Comparison of minority carrier diffusion length dependence on the dose before (after irradiation) and after annealing (left). The recovery of minority carrier diffusion length depends on the dose of gamma-irradiation. Comparison of dependence of calculated activation energy on the dose before and after annealing (right).

SUMMARY AND CONCLUSION

The carrier removal rate in nitrides depends on radiation type, growth method and the dislocation density and distribution. The response of GaN to radiation exposure such as protons can be understood by a simple model involving creation of point defects. Neutron damage creates more extended damage regions. The carrier removal rate by protons, electrons and neutrons in nitride heterostructures increases in the following sequence AlN/GaN>

AlGaN/GaN> InAlN/GaN. This is consistent with their average bond strengths. Proton damage has been the most widely studied and understood. Defects created by protons significantly change the breakdown voltage and off-state stressing response of both InAlN/GaN and AlGaN/GaN HEMTs. The improvement was attributed to the modification of the depletion region due to the introduction of defects. The drain-source currents of both AlGaN/GaN and InAlN/GaN HEMTs decreased 15-20% for proton irradiation with doses of 5 $x10^9$ cm^{-2} and 30-50% decrease at a dose of 5 $x10^{10}$ cm^{-2}. The drain current of proton damaged HEMTs can be partially recovered (40-70%) after annealing at 300°C.

Acknowledgments

The work is supported by an U.S. DOD HDTRA Grant No. 1-11-1-0020, monitored by Dr. James Reed.

References

[1] D.C. Look, D.C. Reynolds, J.W. Hemsky, J.R. Sizelove, R.L. Jones, R.J. Molnar. *Phys. Rev. Lett.* 79, 2273 (1997)

[2] A.Y. Polyakov, N.B. Smirnov, A.V. Govorkov, N.G. Kolin, D.I. Merkurisov, V.M. Boiko, A.V. Korulin, S.J. Pearton, *J. Vac. Sci. Technol. B* 28, 608 (2010)

[3] B.D. White, M. Bataiev, S.H. Gross, X. Hu, A. Karmarkar, D.M. Fleetwood, R.D. Schrimpf, W.J. Schaff, L.J. Brillson, *IEEE Trans. Nucl. Sci.* 50, 1934 (2003)

[4] A.P. Karmarkar, B. Jun, D.M. Fleetwood, D.D. Schrimpf, R.A. Weller, B.D. White, L.S. Brillson and U.K. Mishra, *IEEE Trans. Nucl.* Sci. 51, 3801 (2004)

[5] X Hu, A. Karmarkar, J. Bongim, D.M. Fleetwood, R.D. Schrimpf, R.D. Geil, R.A. Weller, B.D. White, M. Bataiev, L.J. Brillson, U.K. Mishra, *IEEE Trans Nucl Sci*, 50, 1791 (2003)

[6] B. Luo, J.W. Johnson, F. Ren, K.K. Alums, C.R. Abernathy, S.J. Pearton, A.M. Dabiran, A.M. Wowchak, C.J. Polley, P.P. Chow, D. Shoenfeld and A.G. Baca, *Appl. Phys. Lett.* 80, 604 (2002)

[7] B. Luo, J. Kim, F. Ren, J.K. Gillespie, R.C. Fitch, J. Sewell, R. Dettmer, G.D. Via, A. Crespo, T.J. Jenkins, B.P. Gila, A.H. Onstine, K.K. Allums, C.R. Abernathy, S.J. Pearton, R. Dwidevi, T.N. Fogarty, R. Wilkins, *Appl. Phys. Lett.* 82, 1428 (2003)

[8] J. Grant, R. Bates, W. Cunningham, A. Blue, J. Melone, F. McEwan, J. Vaitkus, E. Gaubas, and V. O'Shea, *Nucl. Instrum Meth A* 576, 60 (2007)

[9] Y.S. Puzyrev, T. Roy, E.X. Zhang, D.M. Fleetwood, R.D. Schrimpf, S.T. Pantelides, IEEE Trans. Nucl Sci, 58, 2918 (2011)

[10] C.F. Lo, L. Liu, T.S. Kang, F. Ren, C. Schwarz, E. Flitsiyan, L. Chernyak, H.Y. Kim, J. Kim, S.P. Yun, O. Laboutin, Y. Cao, J.W. Johnson and S.J. Pearton, *J. Vac. Sci. Technol. B* 30, 031202 (2012).

[11] A.Y. Polyakov, S.J. Pearton, P. Frenzer, F. Ren, L. Liu and J. Kim, *J. Mater. Chem. C,* 1, 877 (2013).

[12] S. J. Pearton, R. Deist, F. Ren, L. Liu, A. Y. Polyakov and J. Kim, *J. Vac. Sci. Technol. A* 31, 050801 (2013).

[13] Z. Zhang, A. R. Arehart, E. Cinkilic, J. Chen, E. X. Zhang, D. M. Fleetwood, R. D. Schrimpf, B. McSkimming, J. S. Speck, and S. A. Ringel, *Appl. Phys. Lett.* 103, 042102 (2013).

[14] Y. Berthlet, B. Guhel, H. Boudart, J.L. Gualous, M. Trolet, M. Piccione and C. Gaquiere, *Electronics Lett.,* 48 (2012).

[15] M. R. Hogsed, Y. K. Yeo, M. Ahoujja, M.-Y. Ryu, J. C. Petrosky, and R. L. Hengehold, *Appl. Phys. Lett.* 86, 261906 (2005).

[16] S.A. Vitusevich, N. Klein, A. E. Belyaev, S. V. Danylyuk, M. V. Petrychuk, R. V. Konakova, A. M. Kurakin, A. E. Rengevich, A. Yu. Avksentyev, B. A. Danilchenko, V. Tilak, J. Smart, A. Vertiatchikh and L. F. Eastman, *Phys. Stat. Solidi A* 195, 101 (2003).

[17] A.M. Kurakin, S. A. Vitusevich, S.V. Danylyuk, H. Hardtdegen, N. Klein, Z. Bougrioua, B. A. Danilchenko, R. V. Danilchenko, R. V. Konakova, and A. E. Belyaev, *J. Appl. Phys.,* 103, 083707 (2008).

[18] K. H. Chow, L. Vlasenko, P. Johannesen, C. Bozdog, G. Watkins, A. Usui, H. Sunakawa, C. Sasaoka, and M. Mizuta, *Phys. Rev. B* 69, 045207 (2004).

[19] K. Saarinen, T. Suski, I. Grzegory, and D. C. Look, *Phys. Rev. B* 64, 233201 (2001).

[20] H.Y. Xiao, F. Gao, X. T. Zu, and J. Weber, *J. Appl. Phys.*105, 123527 (2009).

[21] D. C. Look, G. C. Farlow, P. J. Drevinsky, D. F. Bliss, and J. R. Sizelove, *Appl. Phys. Lett.* 83, 3525 (2003).

[22] H. J. von Bardeleben, J. L. Cantin, U. Gerstmann, A. Scholle, S. Greulich-Weber, E. Rauls, M. Landmann, W. G. Schmidt, A. Gentils, J. Botsoa, and M. F. Barthe, *Phys. Rev. Lett.* 109, 206402 (2012)

[23] A. Sasikumar, A. R. Arehart, S. Martin-Horcajo, M. F. Romero, Y. Pei, D. Brown, F. Recht, M. A. di Forte-Poisson, F. Calle, M. J. Tadjer, S. Keller, S. P. DenBaars, U. K. Mishra, and S. A. Ringel, *Appl. Phys. Lett.,* 103, 0335091 (2013).

[24] T. J. Anderson, A.D. Koehler, J.D. Greenlee, B.D. Weaver, M.A. Mastro, J.K. Hite, C.R. Eddy, F.J. Kub and K.D. Hobart, *IEEE Electron Dev. Lett.,* 35,826 (2014)

[25] A.D. Koehler, T.J. Anderson, J.K. Hite, B.D. Weaver, M.J. Tadjer, J.D. Greenlee, P. Specht, M. Porter, T.R. Weatherford, K.D. Hobart, F.J. Kub "Investigation of Proton-Irradiated AlGaN/GaN HEMTs on Sapphire, Si, and SiC Substrates" 56th Electronic Materials Conference, June 25-27, 2014, Santa Barbara, CA

[26] A. Sasikumar, A. R. Arehart, S. W. Kaun, J. Chen, E. X. Zhang, D. M. Fleetwood, R. D. Schrimpf, J. S. Speck and S. A. Ringel, "Defects in GaN based transistors", *Proc. SPIE* 8986, Gallium Nitride Materials and Devices IX 89861C, 1 (2014).

[27] H.Y. Xiao, F. Gao, X. T. Zu and J. Weber, *J. Appl. Phys.*105, 123527 (2009).

[28] D. C. Look, Z. Q. Fang, and B. Claflin, *J. Cryst. Growth,* 281, 143 (2005).

[29] C. G. van de Walle and J. Neugebauer, *J. Appl. Phys.* 95, 3851 (2004).

[30] M.A. Reschikov and H. Morkoc, *J. Appl. Phys.* 97, 061301 (2005).

[31] K. H. Chow, G. D. Watkins, A. Usui, and M. Mizuta, Phys. Rev. Lett. 85, 2761 (2000).

[32] B. Luo, J. Kim, F. Ren, J. Gillespie, R. Fitch, J. Sewell, R. Dettmer, G. Via, A. Crespo, T. Jenkins, B. Gila, A. Onstine, K. Allums, C.R. Abernathy, S.J. Pearton, R. Dwivedi, T. Fogarty and R. Wilkins, *Appl. Phys. Lett.* 82, 1428 (2003).

[33] B.S. Kang, S. Kim, F. Ren, J.W. Johnson, R. Therrien, P. Rajagopal, J. Roberts, E. Piner, K.J. Linthicum, S.N.G. Chu, K. Baik, B.P. Gila, C.R. Abernathy and S.J. Pearton, *Appl. Phys. Lett.* 85, 2962 (2004)

[34] R. Mehandru, B. Luo, J. Kim, F. Ren, B. Gila, A.H. Onstine, C.R. Abernathy, S.J. Pearton, D. Gotthold, R. Birkhahn, B. Peres, R. Fitch, J. Gillespie, T. Jenkins, J. Sewell, D. Via and A. Crespo, *Appl. Phys. Lett.* 82, 2530 (2003).

[35] Y.-S. Hwang, L. Liu, F. Ren, A. Y. Polyakov, N. B. Smirnov, A. V. Govorkov, E. A. Kozhukhova, N.G. Kolin, V. M. Boiko, S S Vereyovkin, V Ermakov, C. F. Lo, O.A. Laboutin, Yu Cao, J. W. Johnson, N. I. Kargin, R. V. Ryzhuk and S.J. Pearton, *J.Vac. Sci.Technol.* B 31, 022206 (2013)

[36] Y.-H. Hwang, S. Li, Y.-L. Hsieh, F. Ren, S. J. Pearton, E. Patrick, M. E. Law and D. J. Smith, *Appl. Phys. Lett.,* 104, 082106 (2014).

[37] B. D. Weaver, P.A. Martin, J.B. Boos and C.D. Cress, *IEEE Trans. Nucl. Sci.,* 59, 3077 (2012).

[38] T. Roy, E. X Zhang. Y.S. Puzyrev, D.M. Fleetwood, R.D. Schrimpf, B.K. Choi, A.B. Hmelo, Anthony and S.T. Pantelides, *IEEE Trans. Nucl. Sci.,*57, 3060 (2010).

[39] L. Selvaraj, T. Suzue, and T. Egawa, *IEEE Electron. Device Lett.,* 30, 587 (2009).

[40] L. Bin and T. Palacios, *IEEE Electron Device Lett.,* 31, 9 (2010).

[41] M. Kanamura, T. Ohki, T. Kikkawa, K. Imanishi, T. Imada, A. Yamada and N. Hara, *IEEE Electron Device Lett.* 31, 189 (2010).

[42] H.-Y. Liu, B.-Y. Chou, W.-C. Hsu, C.-S. Lee, J.-K. Sheu, and C.-S. Ho, *IEEE Trans. Electron. Dev.* 60, 213 (2013).

[43] P. D. Ye, B. Yang, K. Ng, J. Bude, G. Wilk, S. Halder and J. Hwang, *Appl. Phys. Lett.* 86, 063501 (2005).

[44] A.Y. Polyakov, N. Smirnov, A. Govorkov, E. Kozhukhova, S.J. Pearton, F. Ren, L. Liu, J.W. Johnson, W. Lim, N. Kolin, S. Veryokin and V.S. Ermakov, *J. Vac. Sci. Technol. B.* 30, 061207 (2012).

[45] S. Mukherjee, J. Chen, T. Roy, M. Silvestri, R.D. Schrimpf, D.M. Fleetwood, J. Singh, J.M. Hinckley, A. Paccagnella and S.T. Pantelides, *IEEE Trans. Nucl. Sci.*61, 1316 (2014).

[46] Jin Chen, Y.S. Puzyrev, C. X. Zhang, E X Zhang,M.W. McCurdy, D.M. Fleetwood, R.D. Schrimpf, S.T. Pantelides, S.W. Kaun, E.C. Kyle and J.S. Speck, *IEEE Trans. Nucl. Sci.* 60, 4080 (2013).

[47] Ionizing Radiation in MOS Devices/Circuits, ed. T.P. Ma and P.V. Dressdorfer (Wiley, NY, 1989).

[48] F.B. McLean, *IEEE Trans. Nucl. Sci,* 29,1651 (1980).

[49] G. P. Summers, E.A. Burke, P. Shapiro, S.R. Messenger and R.J. Waters, *IEEE Trans Nucl. Sci.,* 40,1372 (1993).

[50] B.R. Tuttle, Defect Interactions of H_2 in SiO_2, “Implications for ELDRS and Latent Interface Trap Buildup," 2010 IEEE Nuclear and Space Radiation Effects Conference, Denver, CO, July 2010.

[51] C. F. Lo, C. Y. Chang, B. H. Chu, H. -Y. Ki, J. Kim, David A. Cullen, Lin Zhou, David. J. Smith, I. I. Kravchenko, S.J. Pearton, Amir Dabiran, B. Cui, and P.P. Chow, S. Jang, and F. Ren, *J. Vac. Sci. Technol. B*. 28, L47 (2010).

[52] K.K. Allums, M. Hlad, A.P. Gerger, B.P. Gila, C.R. Abernathy, S.J. Pearton, F. Ren, R. Dwivedi, T. N. Fogarty and R. Wilkins, *J. Electron Mater.,* 36, 519 (2007).

In: Advances in Applied Physics Research ISBN: 978-1-63483-226-7
Editor: Gabriel Waters

Chapter 2

FLUORESCENCE RESONANCE ENERGY TRANSFER BETWEEN CDTE QUANTUM DOTS AND ORGANIC DYE

***Fang Li*[*]**

Laboratory of Optical Information and Technology, School of Science, Wuhan Institute of Technology,Wuhan, P. R. China

ABSTRACT

In recent years, the light-induced energy transfer process between quantum dots - the organic dye has aroused an interest of many researchers, because quantum dots - organic dye composite system is one of the very promising third-generation photosensitizer and fluorescence biological probe, which has an important significance in the treatment of biomedical tumor. Quantum Dots (QDs) is a kind of novel nano-material. Because of its special size and many special optical properties, it has a wide range of applications in chemical detection, biomedical, electronics. The traditional organic fluorescent dyes, such as Copper phthalocyanine (CuPc) and Rhodamine B (RhB), also have wide applications in the high-technology field due to their special properties of optics, electronics and magnetism. This paper is mainly to study the CdTe quantum dots and organic fluorescent dyes composited system by ultrafast time-resolved spectroscopy technique with two-photon excitation, focusing on the

[*] Corresponding Author address: 693 Xiongchu Avenue, Wuhan, Hubei, P. R. China; Email: lifang_wit@hotmail.com.

fluorescence resonance energy transfer (FRET) of CdTe QDs-CuPc and CdTe QDs-RhB composite system from theory and experiment. In this article, the main contents and conclusions are as follows:

(1) The fluorescence resonance energy transfer in CdTe QDs-CuPc is investigated by ultrafast time-resolved spectroscopy technique equipped with femtosecond laser (780nm,76MHz,130fs). The results show that the fluorescence lifetime of CdTe quantum dots decreases with the increase of CuPc concentration, and the energy transfer efficiency is found to increase with the increase of CuPc concentration. Moreover, the influence of the laser excitation power on the energy transfer efficiency is also studied. It is found that transfer efficiency decreases as excitation laser power increasing. The physical mechanism is the thermal activation with the high power as well as the excited state transitions induced by two-photon excitation. The energy transfer efficiency can reach to be 43.8% when the laser power is 200 mW via two-photon excitation. It indicates that the CdTe QDs-CuPc composite system has high potential as the third generation of photosensitizers.

(2) The fluorescence resonance energy transfer between CdTe quantum dots with different sizes and Rhodamine B (RhB) in aqueous solution is investigated by using the time-resolved fluorescence system under two-photon excitation. With the excitation of 800nm femtosecond laser, result shows that the FRET efficiency of the QDs-RhB system increases with the spectral overlap increase of the CdTe emission spectrum and the Rhodamine B absorption spectrum. Specifically, the fluorescence intensity of QDs decreases and the lifetime of QDs decreases while RhB shows the opposite tendency. By means of the Förster Theory of Energy Transfer, the spectral overlap integral J(λ), Foster radius R0 and the FRET efficiency E are calculated and The FRET characteristics of QD-RhB system is characterized. Theoretical analysis reveals that the physical source is the increase of the sample's Forster radius. Moreover, the relationship of the ratio of acceptor/donor concentration and the FRET efficiency is investigated experimentally. When the ratio of acceptor/donor concentration increases, the lifetime of QDs gets shorter, and the FRET efficiency of the QDs-RhB system becomes higher. The two-photon excited FRET efficiency can reach to 40.1% when the concentration of RhB is 3.0×10^{-5}mol/L. It shows that there is a bright future for these materials in bio-probing, bio-imaging, bio-labelling and other biological and optoelectronic applications.

Keywords: CdTe QDs, CuPc, rhodamine B, fluorescence resonance energy transfer, fluorescence spectroscopy, fluorescence lifetime, two-photon excitation

1. INTRODUCTION

As technology progresses, the desire to increase the performance and functionality of devices while reducing their size is seemingly at odds with one another. Because of the necessity for highly efficient materials that are both durable and small, the field of nanotechnology was born. It was Richard Feynman, in his 1959 Nobel Prize talk [1], who first put forward the way of "bottom up" to preparing nano-scale materials and viewed through the lens of an electron microscope. Today, nano-scale materials are being investigated worldwide and the physical properties of these materials are being explored. The versatility of these structures arises from the size dependent properties they possess that deviate substantially from molecular and bulk [2-4].

While currently the full potential of these materials has yet to be realized, progress in the field has been substantial. Robust methods and finely tuned experimental procedures have created new exotic states of matter such as quantum gases, and semi-metals [5-8]. For semiconductor material, these radical changes occur when quantum-size phenomena take over conventional bulk properties. As the material's size is reduced in at least one dimension (in the 1 to 10nm range), new material's properties start to emerge. These mesoscopic semiconductor structures of reduced dimensions (2-D quantum wells, 1-D quantum wires, and 0-D quantum dots) show unique size-dependent optical and electrical properties due to quantum confinement [9].

The luminescent quantum dots (QDs) are semiconductor nanoparticles (NPs) comprising elements from the periodic groups II-VI or III-V (Table 1). They emit different wavelengths over a broad range of the light spectrum from visible to infrared, depending on their sizes and chemical compositions [10-13]. Most attention focused on the CdSe QDs for emitting fluorescence range 400 to 700 nm and CdTe QDs for emitting fluorescence range near infrared (Figure 1). The size of QDs is ranging between 2 and 10 nm in diameter, which is close to or smaller than the dimensions of the exciton Bohr radius. As a result, the mobility of charge carriers (electrons and holes) is restricted within the nanoscale dimensions, and this quantum confinement effect endows QDs with unique optical and electronic features [14].

Table 1. The type of Quantum Dots

Type	Quantum Dots					
II-VI	MgS	MgSe	MgTe	CaS	CaSe	CaTe
	SrS	SrSe	SrTe	BaS	BaSe	
III-V	BaTe	ZnS	ZnSe	ZnTe	CdS	CdSe
	CdTe	HgS	HgSe	GaAs	InGaAs	InP

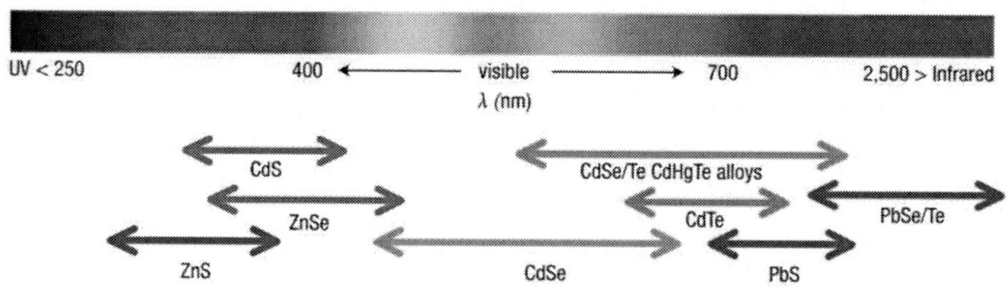

Figure 1. Representative QDs materials scaled as a function of their emission wavelength superimposed over the spectrum [15].

Compared with the traditional organic fluorophores (e.g., organic dyes and fluorescent protein), QDs process several advantages in fluorescence properties for biological applications as following: (1) QDs have broad excitation spectra together with narrow and symmetric emission spectra. Under the same excitation light, different QDs can simultaneously emit different colors, so the multicolor QD probes can be used to image and track multiple molecular targets simultaneously; (2) QDs have very large molar extinction coefficients in the order of $0.5–5.0 \times 10^6$ $M^{-1}cm^{-1}$, which is nearly10–50 times larger than that of organic dyes [16]. Therefore, QDs are able to absorb 10-50 times more photons than organic dyes at the same excitation photon flux, leading to a significant improvement in the probe brightness; individual QDs have been found to be 10–20 times brighter than organic dyes [17]; (3) the maximum emission wavelength of QDs can be controlled in a relatively simple manner by variation of particle size and composition, or through variation of surface coatings. It can be tuned at a variety of precise wavelengths from ultraviolet (UV) to near infrared (NIR). The emission of several QDs like CdHgTe, CdTeSe/CdS can be engineered in the range of 700–900 nm, which is a “transparent window” of biological species that can effectively eliminate the light absorbance and background interference of tissues and body fluids (Figure 2). Thus NIR QDs have found broad applications in living animal fluorescence imaging; (4) The stability with regard to photobleaching is much more superior to conventional fluorophores, which enable QDs to be applied for highly sensitive detection and long observation times in fluorescence microscopy.

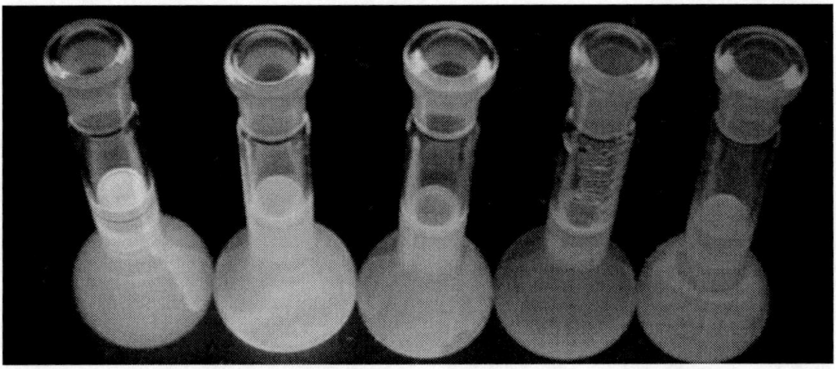

Figure 2. Photo demonstrating the size-tunable fluorescence properties and spectral range of the five CdTe QD dispersions. All samples were excited at 400nm.

In addition, QDs are nanocrystals composed of a core of a semiconductor, usually composed of elements from groups II–IV, e.g., CdSe, or groups III–V, e.g., InP. The shell is typically a higher bandgap material such as ZnS. Finally, a capping outer layer such as silica can offer large-surface area for covalently linking to biorecognition molecules such as peptides, antibodies, nucleic acids and small-molecule ligands for further applications [18] (Figure 3).

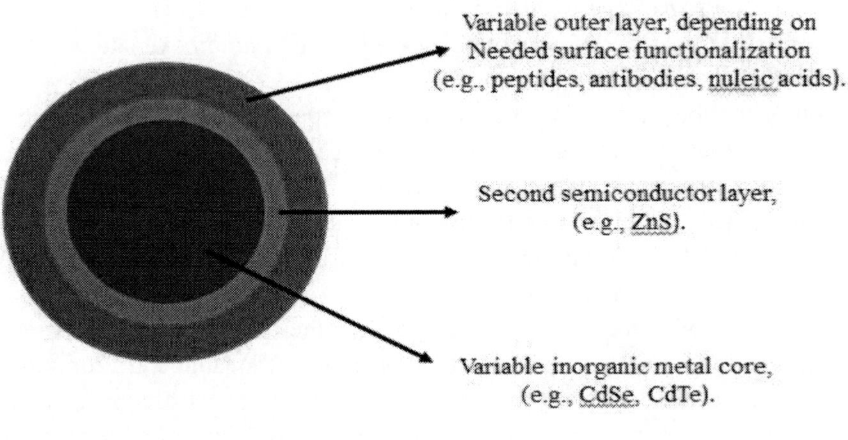

Figure 3. Schematic representation of a quantum dot.

The prosperity that QDs achieved in biolabeling inspired pharmaceutical researchers to apply them to the research and development of novel nanomedicines. There are two main aspects for the application of QDs in this field. One is to develop fluorescent diagnostic nanoprobes for cancer detection and therapy after modification of certain targeting molecules. A single QD is large enough for conjugation to multiple ligands, leading to enhanced binding affinity to the targets. Together with their brightness and anti-bleaching advantage, the functionalized QDs can be used to capture and quantify a panel of biomarkers on intact cancer cells and tissue specimens sensitively and specifically, allowing a correlation of traditional histopathology and molecular signatures for the same material. The second application of QDs in nanomedicine is to label drug molecules or nanocarriers. With the aid of sensitive and fast responsive fluorescent imaging system that has been developing into a promising platform for drug investigation and screening, we can expect to acquire real-time information on distribution, transportation, drug release and pharmacodynamic mechanism of nanomedicine in biospecimens intuitively.

The photoinduced energy transfer in the QD-phthalocyanine system were usually characterized using UV-visible lasers with a wavelength less than 700nm, on the basis of one-photon absorption (OPA) process [19-21]. To our knowledge, few studies have been devoted to two-photon absorption (TPA) process. In contrast to the OPA photodynamic therapy (PDT), TPA-PDT attracts more attentions since it has two main advantages. Firstly, the near-infrared wavelength (700-1100nm) used in two-photon excitation PDT meets the optical transparent window of mammalian tissue, which reduced light absorption and scattering by tissue observed in one photon excitation PDT so that it allows deeper examination and treatment of tumor. Secondly, two-photon excitation PDT possesses a quadratic dependence on laser intensity so that the illumination area is confined in the focal volume, which enhances the spatial resolution and avoids the unwanted damage to the healthy tissue. Therefore, on further research, the ultrafast nonlinear response is quite essential for understanding the source of NLA in the femtosecond regime and their practical applications [22].

In this work, the FRET efficiency in the CdTe QDs and organic fluorescent dyes (such as copper phthalocyanine, Rhodamine B) composited system were investigated by ultrafast time-resolved spectroscopy technique equipped with femtosecond laser. On the other hand, in order to reveal the source of the FRET efficiency, an efficient energy level model was carried out. Our studies show that the CdTe QDs and organic fluorescent dyes composited system is a promising candidate in two-photon-excitation PDT for third-generation photosensitizer and fluorescence biological probe.

2. THEORY

Fluorescence is the emission of light by a substance that has absorbed light or other electromagnetic. It is a form of luminescence. In most cases, the emitted light has a longer wavelength, and therefore lower energy, than the absorbed radiation. The most striking examples of fluorescence occur when the absorbed radiation is in the ultraviolet region of the spectrum, and thus invisible to the human eye, and the emitted light is in the visible region.

Fluorescence analysis method caused a great number of attentions since it was born, and it was developed rapidly and applied widely. In addition to this method itself has a high-sensitivity advantages, it is also propitious to time scale. Fluorescence resonance energy transfer (FRET) has become widely used in all applications of fluorescence, including medical diagnostics, DNA analysis, and optical imaging [23-27]. The widespread use of FRET is due to the favorable distances for energy transfer, which are typically the size of a protein or the thickness of a membrane. Additionally, the extent of FRET is readily predictable from the spectral properties of the fluorophores. These favorable properties allow for the design of experiments based on the known sizes and structural features of the sample.

1. History of Fluorescence Analysis

An early observation of fluorescence was described in 1565 by N. Monardes in the fusion known as lignum nephriticum. It was derived from the wood of two tree species [28, 29]. In 1819, Edward D. Clarke [30] and in 1822 Rene Just Hauy [31] described fluorescence in fluorites, Sir David Brewster described the phenomenon for chlorophyll in 1833 [32] and Sir John Herschel did the same for quinine in 1845 [33]. There had been little progress in explaining the fluorescence phenomena while some other fluorescent materials and solutions were found during the 17th and 18th centuries, ever since the fluorescence phenomena was first observed by George Gabriel Stokes in 1852 at the University of Cambrige. These early experiments used relatively simple instrumentation, and the ability of fluorspar and uranium glass to change invisible light beyond the violet end of the visible spectrum into blue light. He named this phenomenon "fluorescence," as it was derived from the mineral fluorite, some of which contain traces of divalent europium, which serves as the fluorescent activator to emit blue light. Then he observed the inner filter effect and suggested using optical properties such as absorption, colored reflection, and fluorescence to identify organic substances [34]. Moreover, he also proposed to use fluorescence analysis firstly. It was not until 1867 that Goppelsröder who did work through fluorescence analysis for the first time, determining the aluminum in aluminium-morin complexes [35]. Sir Liebeman put forward the rule for studying the relationship between immunochemical structure and fluorescence in 1880 [36]. By the end of the 19th century, people have already known hundreds of fluorescent chemicals, such as fluorescein, eosin, polycyclic aromatic hydrocarbon, etc.

Since the 20th century, more research about fluorescence was carried out: Sir Wood discovered resonance fluorescence in 1905 [37]; Professor Frank and Hertz developed a quantitative investigation in 1914 [38], using electron bombardment; Then sensitive fluorescence was found by Sir Frank and Cario in 1922 [39]; Professor Wawillow measured the absolute fluorescent yield in 1924 [40]; Then the direct measurement of lifetime was promoted by Gaviola in 1926 [41].

During the past 20 years there has been a remarkable growth in the use of fluorescence in the biological sciences. fluorescence analysis and time-resolved fluorescence are considered to be primarily research tools in biochemistry and biophysics.

This emphasis has changed, and the use of fluorescence has expanded. fluorescence analysis can be applied to a wide range of problems in the chemical and biological sciences. The measurements can provide information on a wide range of molecular processes, including the interactions of solvent molecules with fluorophores, rotational diffusion of biomolecules, distances between sites on biomolecules, conformational changes, and binding interactions.

The usefulness of fluorescence is being expanded by advances in technology for cellular imaging and single-molecule detection. These advances in fluorescence technology are decreasing the cost and complexity of previously complex instruments. Fluorescence analysis will continue to contribute to rapid advances in biology, biotechnology and nanotechnology.

2. Emission and Absorption Spectra

Fluorescence is a photoluminescence phenomenon, which is due to the selective absorption of light from molecules, and the different incident wavelengths lead to different excitation efficiency. Fixing emission wavelengths (i.e., determination wavelengths) while changing the excitation wavelengths, for the sake of recording the corresponding fluorescence intensity, then we obtain the excitation spectrum, which the fluorescence intensity associates with the excitation wavelength. In addition, fixing excitation wavelengths while changing the emission wavelengths, in order to record the corresponding fluorescence intensity, then we obtain the emission spectrum, which the fluorescence intensity associates with the emission wavelength.

In physics, fluorescence spectral data are generally presented as emission spectra. It is the process by which a higher energy quantum mechanical state of a particle becomes converted to a lower one through the emission of a photon, resulting in the production of light. The frequency of light emitted is a function of the energy of the transition. Since energy must be conserved, the energy difference between the two states equals the energy carried off by the photon. The energy states of the transitions can lead to emissions over a very large range of frequencies.

The generally symmetric nature of these spectra is a result of the same transitions being involved in both absorption and emission, and the similar vibrational energy levels of S_0 and S_1. In most fluorophores, these energy levels are not significantly altered by the different electronic distributions of S_0 and S_1. Suppose the absorption spectrum of a fluorophore shows distinct peaks due to the vibrational energy levels. Such peaks are seen for anthracene in Figure 4. These peaks are due to transitions from the lowest vibrational level of the S_0 state to higher vibrational levels of the S_1 state. Upon return to the S_0 state the fluorophore can return to any of the ground state vibrational levels. These vibrational energy levels have similar spacing to those in the S_1 state. The emission spectrum shows the same vibrational energy spacing as the absorption spectrum.

Fluorescence spectral data are generally presented as emission spectra. The fluorescence emission spectrum is a plot of intensity versus wavelength (nanometer) or wavenumber (cm-1). Two typical fluorescence emission spectra are shown in Figure 4. Emission spectra vary widely and are dependent upon the chemical structure of the fluorophore and the solvent in which it is dissolved. The spectra of some compounds, such as perylene, show significant structure due to the individual vibrational energy levels of the ground and excited states. Other compounds, such as quinine, show spectra devoid of vibrational structure.

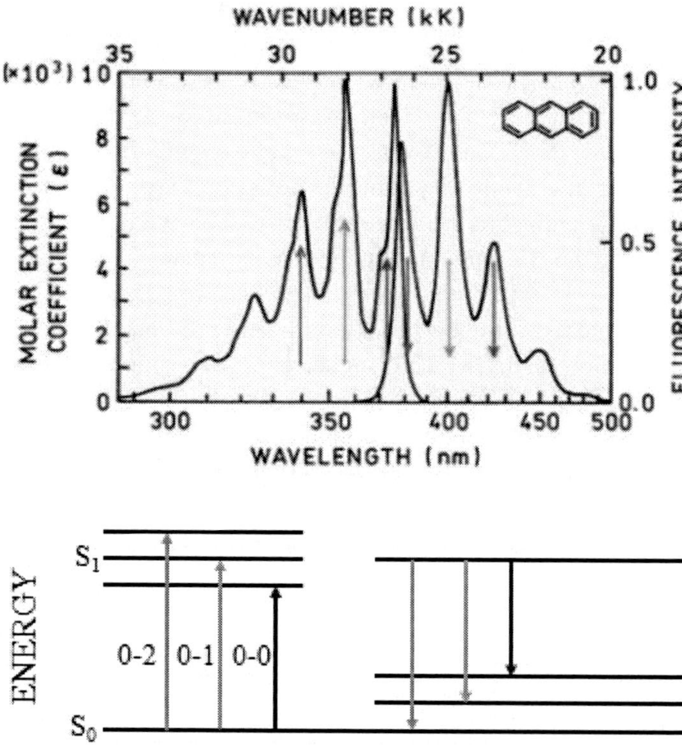

Figure 4. Mirror-image rule and Frank-Condon factors. The absorption and emission spectra are for anthracene. The numbers 0,1, and 2 refer to vibrational energy levels [42].

For most substances, the amount of emission varies with the temperature and the spectroscopic composition of the object, leading to the appearance of color and emission. Precise measurements at many wavelengths allow the identification of a substance via emission spectroscopy.

Professor Jablonski explained the experimental facts he distinguished the experimental facts he distinguished the transition moments in absorption and in emission and analyzed various factors responsible for the depolarization of luminescence [43, 44]. A typical Jablonski diagram is shown in Figure 5. The singlet ground, first, and second electronic states are depicted by S_0, S_1, and

S_2, respectively. At each of these electronic energy levels the fluorophores can exist in a number of vibrational energy levels, depicted by 0, 1, 2, etc. The transitions between states are depicted as vertical lines to illustrate the instantaneous nature of light absorption. Transitions occur in about 10^{-15}s, a time too short for significant displacement of nuclei. This is the Franck-Condon principle.

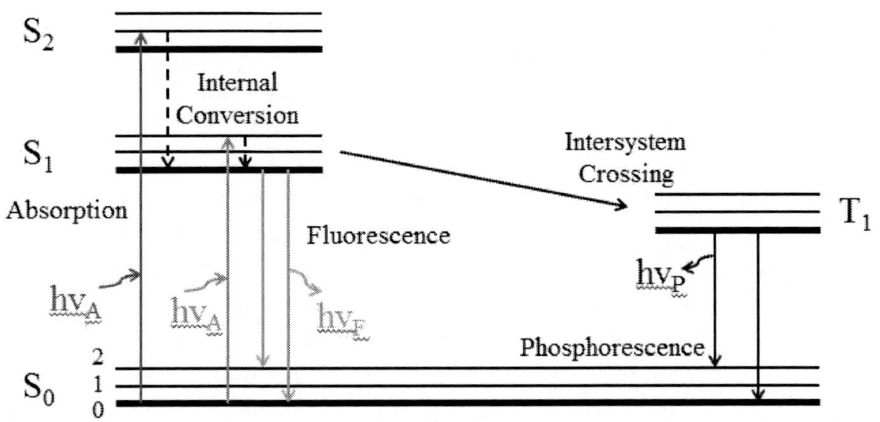

Figure 5. One form of a Jablonski diagram.

The energy spacing between the various vibrational energy levels is illustrated by the emission spectrum of perylene (Figure 4). The individual emission maxima (and hence vibrational energy levels) are about 1500 cm^{-1} apart. At room temperature thermal energy is not adequate to significantly populate the excited vibrational states. Absorption and emission occur mostly from molecules with the lowest vibrational energy. The larger energy difference between the S_0 and S_1 excited states is too large for thermal population of S_1. For this reason we use light and not heat to induce fluorescence.

Following light absorption, several processes usually occur. A fluorophore is usually excited to some higher vibrational level of either S_1 or S_2. With a few rare exceptions, molecules in condensed phases rapidly relax to the lowest vibrational level of S_1. This process is called internal conversion and generally occurs within 10^{-12} s or less. Since fluorescence lifetimes are typically near 10^{-8} s, internal conversion is generally complete prior to emission. Hence, fluorescence emission generally results from a thermally equilibrated excited state, that is, the lowest energy vibrational state of S_1.

Return to the ground state typically occurs to a higher excited vibrational ground state level, which then quickly (10^{-12} s) reaches thermal equilibrium (Figure 5). Return to an excited vibrational state at the level of the S_0 state is the reason for the vibrational structure in the emission spectrum of perylene. An interesting consequence of emission to higher vibrational ground states is that the emission spectrum is typically a mirror image of the absorption spectrum of the $S_0 \rightarrow S_1$ transition. This similarity occurs because electronic excitation does not greatly alter the nuclear geometry. Hence the spacing of the vibrational energy levels of the excited states is similar to that of the ground state. As a result, the vibrational structures seen in the absorption and the emission spectra are similar.

Molecules in the S_1 state can also undergo a spin conversion to the first triplet state T_1. Emission from T_1 is termed phosphorescence, and is generally shifted to longer wavelengths (lower energy) relative to the fluorescence. Conversion of S_1 to T_1 is called intersystem crossing. Transition from T_1 to the singlet ground state is forbidden, and as a result the rate constants for triplet emission are several orders of magnitude smaller than those for fluorescence. Molecules containing heavy atoms such as bromine and iodine are frequently phosphorescent. The heavy atoms facilitate intersystem crossing and thus enhance phosphorescence quantum yields.

3. Steady-State Absorption Theory

In the steady-state absorption method, if the substance absorbed a light which was emitted radiation, that is so-called fluorescence. The intensity of fluorescence (I_f) was related to the absorption intensity (I_a) and the fluorescence quantum yield (Y_f) , namely,

$$I_f = Y_f I_a \tag{1}$$

where I_f is equal to the transmission intensity minus the incident intensity, we obtain:

$$I_f = Y_f(I_0 - I_t) = Y_f I_0 (1 - {}^{I_t}/_{I_0}) \tag{2}$$

By the law of Lambert-Beer, $I_t/I_0 = e^{-abc}$, we obtain:

$$I_f = Y_f I_0 (1 - e^{-abc}) \tag{3}$$

We study the function e^{-abc} by Taylor, we have

$$e^{-abc} = 1 - abc + (abc)^2/2! - (abc)^3/3! + \cdots \tag{4}$$

While abc is close 0, we have $e^{-abc} = 1 - abc$, taking to Eq.(3),

$$I_f = Y_f I_0 abc \tag{5}$$

Using the molar absorption coefficient ε take place a, we obtain:

$$I_f = 2.303 Y_f I_0 \varepsilon bc \tag{6}$$

We may find that, for some dilute fluorescent material solution, the fluorescence intensity is proportional to the concentration of the substance, while the frequency and intensity of laser is at certain.

4. Fluorescence Resonance Energy Transfer

Fluorescence resonance energy transfer is a process that involves nonradiative energy transfer from a photoexcited donor molecule, after absorption of a higher energy photon, to an acceptor molecule of a different species (brought in close proximity), which may relax to its ground state by emitting a lower energy photon. This process results from dipole-dipole interactions and is thus strongly dependent on the center-to-center separation distance ; it requires a nonzero integral of the spectral overlap between donor emission and acceptor absorption [45]. Careful selection of an appropriate donor-acceptor pair ensures high transfer efficiency and provides two measurable parameters: a quenched donor photoemission and enhanced acceptor fluorescence. While good spectral overlap is paramount to high transfer efficiency, the donor and acceptor photoluminescence (PL) signals must also be well resolved to extract accurate experimental information for the system under investigation, e.g., a binding event due to a specific interaction in a biological receptor-ligand system or a change in protein conformation due to interactions with a target molecule. In addition, the excitation line should be chosen to coincide with the minimum of the acceptor absorption spectrum in order to reduce contributions resulting from direct excitation of the acceptor. Satisfying all of these conditions simultaneously is inherently difficult using conventional organic dyes with narrow excitation spectra and small achievable Stokes shifts. The rate of energy transfer for an isolated single D-A pair separated by a distance r can be expressed using the Forster formalism as [46],

$$K_{D-A} = \frac{B \times Q_D I}{\tau_D r^6} = \left(\frac{1}{\tau_D}\right) \times \left(\frac{R_0}{r}\right)^6 \tag{7}$$

where Q_D is the quantum yield of the donor and τ_D is the excited-state lifetime of the donor; the constant B is a function of the refractive index of the medium nD, Avogadro's number N_A, and a parameter, κp, that depends on the relative orientation of the donor and acceptor dipoles:

$$B = \frac{[9000 \times (ln10)]\kappa_p^2}{128\pi^5 n_D^4 N_A} \tag{8}$$

where $\kappa_p^2 = 2/3$ for randomly oriented dipoles and varies between 0 and 4

for the cases of orthogonal and parallel dipoles, respectively. The overlap integral, I, defined as

$$I = \int_0^{\infty} PL_{D-corr}(\lambda)\, \epsilon_A(\lambda) \lambda^4 d\lambda \tag{9}$$

is a quantitative measure of the donor-acceptor spectral overlap over all wavelengths λ; it is a function of the normalized donor emission spectrum (dimensionless property), $PL_{D\text{-}corr}$, and the acceptor absorption spectrum (expressed as an extinction coefficient),. In Eq(7), we introduced the Forster radius (or distance), R_0, which is defined as

$$R_0 = (BQ_D I)^{1/6} = \left(\frac{9000(ln10)\kappa_p^2 Q_D}{N_A 128\pi^5 n_D^4} I\right)^{1/6} \quad (10)$$

and corresponds to a separation distance at which the rate of transfer matches the rate of exciton decay: $k_{D-A} = \tau_D^{-1}$. Note that the quantum yield must be measured to give an accurate estimate of the Forster distance. The FRET efficiency, E, defined as

$$E = \frac{k_{D-A}}{k_{D-A} + \tau_D^{-1}} = \frac{R_0^6}{R_0^6 + r^6} \quad (11)$$

accounts for the fraction of excitons that are transferred from the donor to the acceptor nonradiatively. Using Eq(11) for the FRET efficiency, one can also define the Forster distance, R_0, as the separation distance that yields 50% energy transfer efficiency. In the case where one donor species can interact with several acceptors brought in close proximity simultaneously, the above analysis can be modified to account for the presence of these complex interactions, and the above efficiency can be expressed as

$$E = \frac{nR_0^6}{nR_0^6 + r^6} \quad (12)$$

where n is the average number of acceptor molecules interacting with one donor. In the above configuration the presence of several acceptor fluorophores increases the rate of FRET. It may be described as an effective linear enhancement of the energy transfer cross section between a donor and several acceptors arrayed around its center.

The FRET efficiency can be measured experimentally and is commonly defined as

$$E = 1 - \frac{F_{DA}}{F_D} \quad (13)$$

where F_{DA} is the integrated fluorescence intensity of the donor in the presence of the acceptor(s) and F_D is the integrated fluorescence intensity of the donor alone (no acceptors present).

For FRET within the quasi-monodispersed QD assembly, where these QDs act as donors, the FRET efficiency can be estimated as

$$E = 1 - \frac{\tau_{DA}}{\tau_D} \tag{14}$$

where τ_D and τ_{DA} are the lifetime of the donor in the absence and presence of the acceptor [47]. The Forster theory treats the donor and acceptor as points in the interaction space, whereas the nanocrystals have finite size and are relatively large compared to the dye molecules.

3. Method

Time-resolved spectroscopy is defined as technology for detecting the temporal dynamics in the photophysical process, the absorption, excitation, emission process and some other information of materials can be obtained through this technique, which is essential for researching and developing the photoelectric functional materials. With the development of laser-pulse technology, the time-resolved resolution improves constantly. The time-resolved spectroscopy, based on femtosecond pulse laser, has become one of the most effective tools for developing new types of materials and investigating the various intramolecular ultrafast dynamics process, it is beneficial to development of the photoelectric conversion, photochemical catalysis materials, what's more, the photodynamic drugs, it is also a powerful tool for exploring the substances' photophysical and photochemical properties in a large number of fields. Main methods to measure the materials focusing on time-resolved measurement are: Femtosecond pump-probe technique (Transient absorption spectra, TA), femtosecond fluorescence up-conversion technique (UC) and time-correlated single photon counting (TCSPC).

1. Femtosecond Pump-Probe Technique (Transient Absorption Spectra)

Transient absorption spectra technique is a useful tool for exploring the excited state and intermediate product during the photochemical process, qualitatively and quantitatively. And the transient absorption system, which concentrates on the photodynamic and photochemical process through studying the excited states and the short lifetime intermediates caused by incident laser, its contributions are in two aspects, that is, research on excited states and intermediates in chemical reactions and energy transfer in excited states, the former contains characterization of excited state and intermediate product, along with reaction kinetics, the other consists of investigation on excitation energy transfer, electron transfer and proton transfer. The main characteristics of transient absorption spectra are as follows: (1) Detecting the transient absorption spectra obtained from the product which is from incident light absorption, we may find the transient product easier than emission spectra. (2) Real-time tracking the transient product which is due to the detection is time-resolved, leading to the particular suitable to observe the evolution of the excited states and intermediates in the photochemical reaction. Moreover, it is not only propitious to confirm reaction mechanism but also helpful to catch sight of new phenomenon.

A relative delay femtosecond probe pulse to detect the transmission spectroscopy which is excited by another femtosecond pulse, the power of probe laser is weaker than the pump pulse (Figure 6).

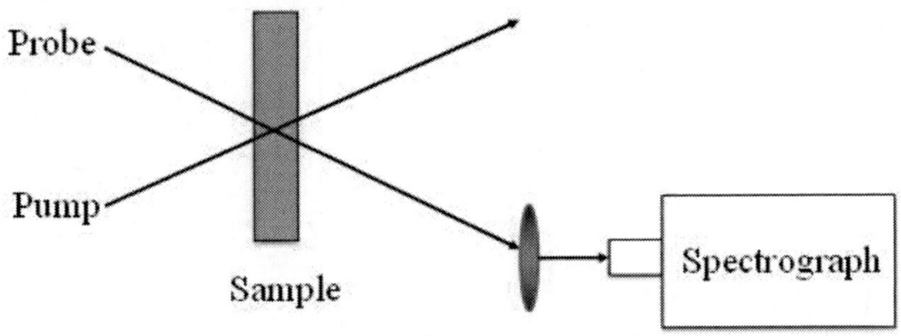

Figure 6. Schematic of Pump-probe technique.

Initially, the detector of pump-probe technique is photodiode, which leads to single wavelength absorption spectrum through each probe, in order to obtain a series of absorption wavelengths, we change the wavelength of monochromator, then the transient absorption spectroscopy about full wave-range start to restructure. With the development of detection technology, charge coupled device (CCD) is coming into being, which makes the multi-wavelength detection possible. What’s more, people use the laser-induced medium, such as sapphire, water, calcium fluoride, etc, which aroused the nonlinear effect of refractive index, then super-continuum white light generates through self-phase modulation, it is the beginning of white light pump-probe technique. Combined with the muti-wavelength detection of CCD, it is convenient for obtaining the full absorption wavelength. This femtosecond transient absorption technique has been widely used in many fields, such as the dynamics of carrier relaxation in semiconductor nano-materials, electron transfer in the process of photoelectric conversion and the dynamics of molecular reaction.

2. Femtosecond Fluorescence Up-Conversion Technique

The ultimate time resolution is provided by methods that bypass the limited time-resolution of the detectors, and rely on the ps and fs pulse widths available with modern lasers. The basic idea is to pass the fluorescence signal through an up-conversion crystal, and to gate the crystal with another ps or fs light pulse (Figure 7).

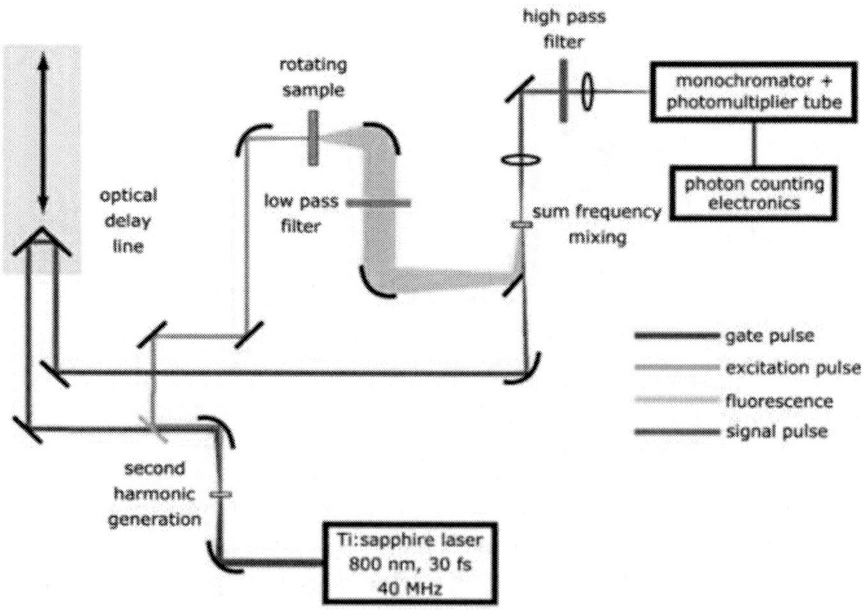

Figure 7. Schematic for fluorescence up-conversion measurements.

One observes the shorter wavelength harmonic generated by the combined effects of the laser pulse and the emission. In this schematic the sample is excited with the second harmonic of a Ti:sapphire laser. The emission is focused on the up-conversion crystal. The fundamental output of the Ti:sapphire is focused on the crystal in the same region where the fluorescence is focused. An up-conversion signal is generated that is proportional to the emission at the moment the gating pulse arrives. The time resolution is determined by the width of the laser pulse. The time-resolved decay is obtained by measuring the intensity of the up-converted signal as the optical delay time is varied. The signals are typically weak, so that an optical chopper and lock-in detectors sometimes are needed to measure the up-converted signal in the presence of considerable background.

Up-conversion provides impressive time resolution; however, the instrumentation is rather complex. For instance, even a seemingly minor change of the emission wavelength can require a major readjustment of the apparatus since the orientation of the up-conversion crystal has to be adjusted according to the wavelength. Decay times of more than 1–2 ns are difficult to measure because of the use of a delay line (≈1 ns/foot). Alignment of the delay line can be quite difficult to maintain as the time delay is altered.

3. Time-Correlated Single Photon Counting (TCSPC)

Time-correlated single photon counting (TCSPC) technique is often used as the measurement of fluorescence lifetime. It has a lot of advantages, such as the shorter measure time, the higher sensitivity and the excellent signal-noise ratio. Moreover, it measures the time of each photon arrives to the detector and obtains the principles of fluorescence decay lifetime. These instruments use high repetition rate mode-locked picosecond (ps) or femtosecond (fs) laser light sources, and high-speed microchannel plate (MCP) photomultiplier tubes (PMTs). For many applications, these expensive systems are being rapidly replaced by systems using pulsed-laser diodes (LDs), light-emitting diodes (LEDs), and small, fast PMTs.

The principle of TCSPC is somewhat unique (Figure 8). The sample is excited with a pulse of light, resulting in the waveform shown at the top of the Figure. This is the wave form that would be observed when many fluorophores are excited and numerous photons are observed. However, for TCSPC the conditions are adjusted so that less than one photon is detected per laser pulse. In fact, the detection rate is typically 1 photon per 100 excitation pulses. The time is measured between the excitation pulse and the observed photon and stored in a histogram. The x-axis is the time difference and the y-axis the number of photons detected for this time difference. When much less than 1 photon is detected per excitation pulse, the histogram represents the waveform of the decay. If the count rate is higher the histogram is biased to shorter times. This is because with TCSPC only the first photon can be observed. At present the electronics are not fast enough to measure multiple photons per pulse when the lifetimes are in the nanosecond range.

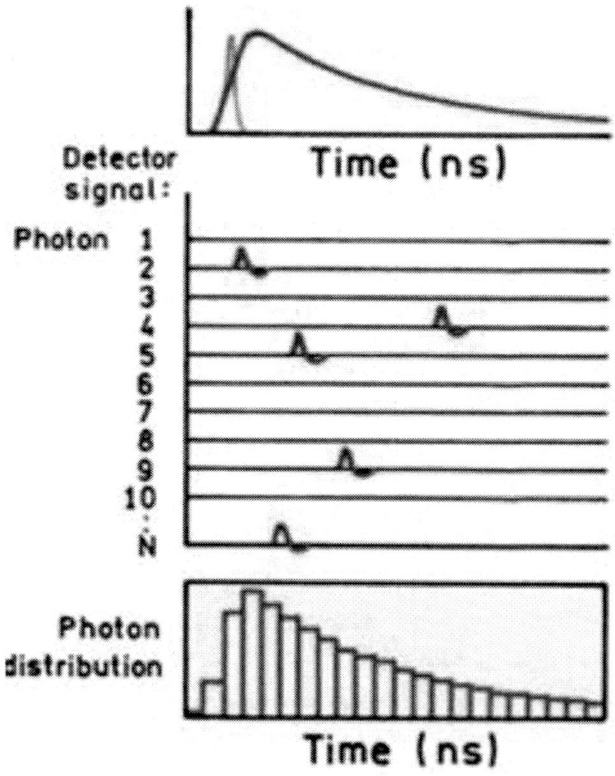

Figure 8. Principle of TCSPC. The pulses in the middle panel represent the output from a constant fraction discriminator.

Multiple photons per pulse can be measured for decay times near a microsecond or longer. Specialized electronics are used for measuring the time delay between the excitation and emission (Figure 9). The experiment starts with the excitation pulse that excites the samples and sends a signal to the electronics. This signal is passed through a constant function discriminator

(CFD), which accurately measures the arrival time of the pulse. This signal is passed to a time-to-amplitude converter (TAC), which generates a voltage ramp that is a voltage that increases linearly with time on the nanosecond timescale. A second channel detects the pulse from the single detected photon. The arrival time of the signal is accurately determined using a CFD, which sends a signal to stop the voltage ramp. The TAC now contains a voltage proportional to the time delay (Δt) between the excitation and emission signals. As needed the voltage is amplified by a programmable gain amplifier (PGA) and converted to a numerical value by the analog-to-digital converter (ADC). To minimize false readings the signal is restricted to given range of voltages. If the signal is not within this range the event is suppressed by a window discriminator (WD). The voltage is converted to a digital value that is stored as a single event with the measured time delay. A histogram of the decay is measured by repeating this process numerous times with a pulsed-light source. The convolution formula is

$$\mathrm{I} = \int_0^\infty PL_{D-corr}(\lambda)\,\epsilon_A(\lambda)\lambda^4 d\lambda \qquad (15)$$

where is pulse response function,and is the measured signal function.

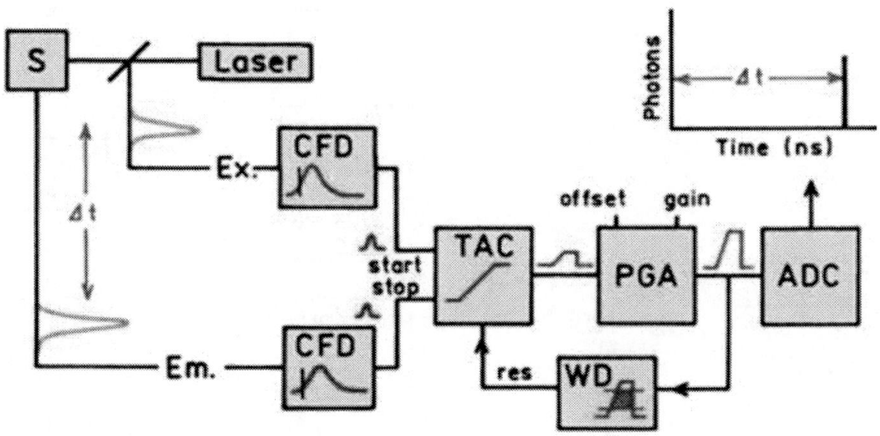

Figure 9. Principle of TCSPC. The pulses in the middle panel represent the output from a constant fraction discriminator.

4. Time-Resolved Fluorescence Test System

In our study, to analyze the kinetics of two-photon excited luminescence, time resolved two-photon photoluminescence (PL) and fluorescence lifetime measurements were performed using the time-resolved fluorescence test system. The samples were excited at a wavelength with femtosecond laser pulses, produced by Titanium sapphire femtosecond laser with pulse widths of 130fs, a repetition frequency of 76MHz, and fixed power of 500mw.

The laser passes through aperture slot after reflecting by reflector mirror, then goes through halfwave plate and glan prism after the light elimination by aperture, where the halfwave plate is used for polarization while glan prism is used for detection of polarization, both of them are used for adjusting optical powers. Next, the laser which is from glan prim will go through the reflector and lens1, and the fluorescence spectra were measured by the fluorescence spectrometer and the fluorescence decay curves were recorded using a single photon counter. The samples were placed in a 10-mm cuvette, and to avoid the interference from the excitation source, the cuvette installed at 45 degree angle to allow the spectrometer to collect the reflection fluorescence of the sample. Figure 10 presents a schematic diagram of the time-resolved fluorescence test system.

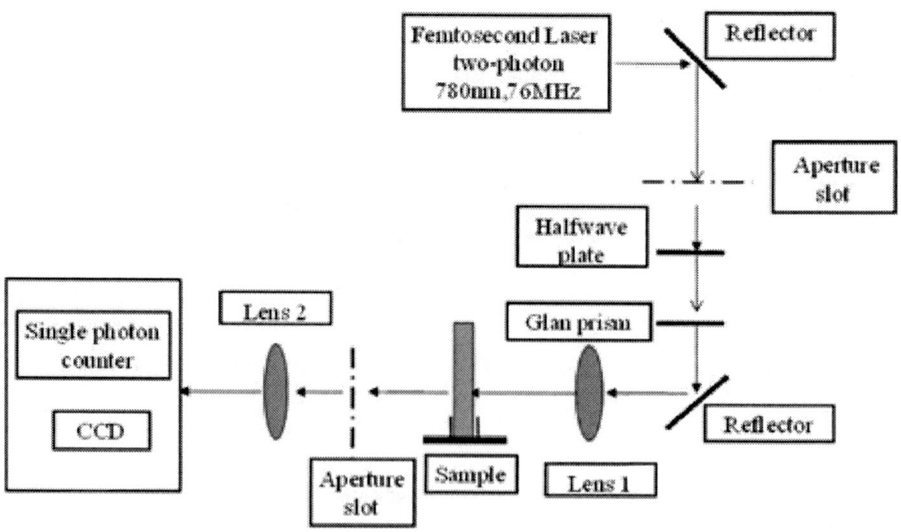

Figure 10. Schematic diagram of the time-resolved fluorescence test system.

4. Fluorescence Resonance Energy Transfer Experiments of CdTe Quantum Dots-Cooper Phthalocyanine Composite System

Phthalocyanine has been growing rapidly as "functional dyes" because of its special properties of optics, electronics and magnetism in the field of high-tech [48-51]. In recent years, researchers have studied it via using various pulse width and different wave-band, besides, with the rapid development of high intensity femtosecond laser and nonlinear optics, it leads to a number of new areas on the interaction between laser and materials [52-55].

In this part, two-photon excited energy transfer in a non-conjugated system of CdTe quantum dots (QDs)-copper phthalocyanine (CuPc) using time-resolved fluorescence spectra was reported. The effects of laser excitation intensity on the fluorescence lifetime and energy transfer efficiency were discussed. Furthermore, the physical mechanism of energy transfer in the QDs-phthalocyanine system under two-photon excitation was discussed.

4.1. Optical Experiments of CdTe QDs-CuPc Composite System

4.1.1. Experimental Materials

CdTe QDs were prepared from Cd^{2+} and NaHTe in the presence of thioglycolic acid in aqueous solution by electrostatic reaction. Figure 11(a) is the TEM image of CdTe nanoparticles, showing the average size is about 3.5nm. CuPc are provided by Shanghai Pin'ao pigment Co. Ltd. Both CdTe QDs and CuPc are water-soluble, allowing in vivo biological applications. CdTe QDs–CuPc mixtures dissolved in deionized water are prepared at a series of CuPc concentrations with constant stirring for 5 min to ensure that the solution is homogeneous. The concentrations (unit: mol/L) of the two solutions are of the 10^{-6} ~10^{-4} order of magnitude to ensure that there is little reabsorption during the PL measurements. Figure 11(b) is the TEM image of CdTe-CuPc sample, showing the average size is about 70nm. We used cuvette of pathlength of 1mm to place the sample solution, the volume ratio of CdTe quantum dots and CuPc is keeping 1:3.

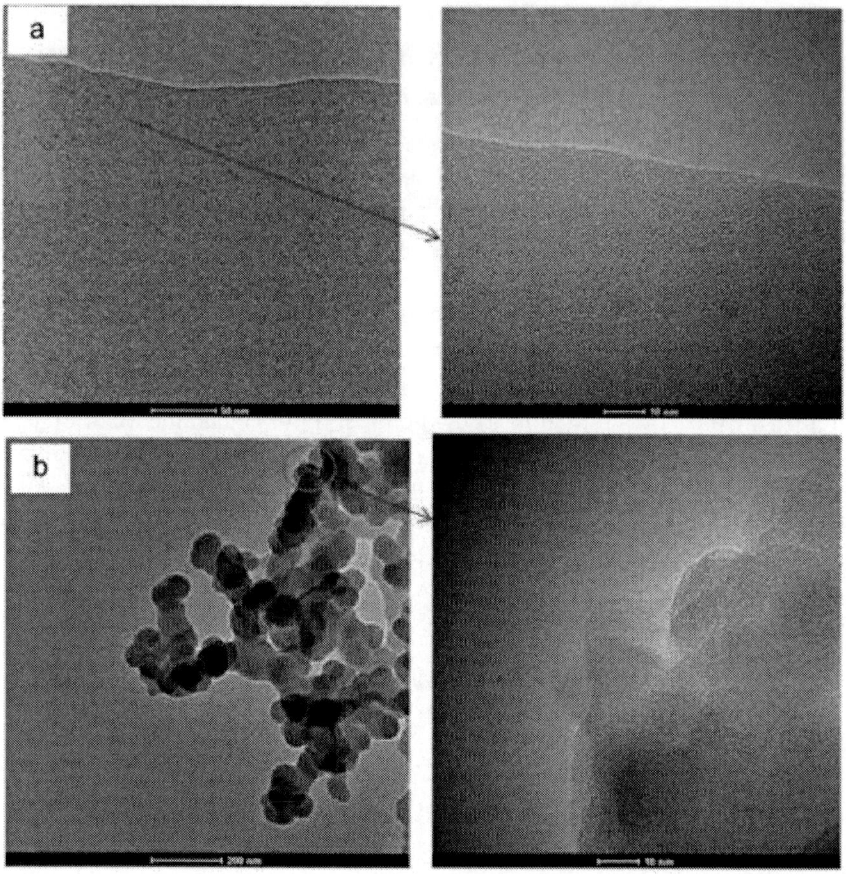

Figure 11. TEM images of (a) CdTe and (b) CdTe-CuPc.

All the absorption spectra are measured with a UV–visible spectrometer (HITACHI U-3310, Japan), and the steady-state fluorescence spectra are measured with a fluorescence spectrometer (Jasco FP-6500, Japan). A time-correlated single photon counting (TCSPC) system consisting of a PicoHarp 300 controller, a PDL 800-B reference, and a SPDA-15 detector is used for the time-resolved fluorescence measurements on a nanosecond time scale. A wavelength of 780 nm from a femtosecond laser system (Mira 900, Coherent, 76 MHz, 130 fs) is used as the two-photon excitation light source. The fluorescence signal is detected with a spectrometer (Andor SR-500i, DU970P-BVEM CCD). The fluorescence decays are fitted by a monoexponential model $I(t) = a_0 + a_1\exp(-t/\tau)$.

4.1.2. The Steady-State Absorption Theory

In the steady-state absorption method, if the substance absorbed a light which was emitted radiation, that is so-called fluorescence. The intensity of fluorescence (I_f) was related to the absorption intensity (I_a) and the fluorescence quantum yield (Y_f) , namely,

$$I_f = Y_f I_a \tag{16}$$

where I_f is equal to the transmission intensity minus the incident intensity, we obtain:

$$I_f = Y_f(I_0 - I_t) = Y_f I_0(1 - {}^{I_t}/_{I_0}) \tag{17}$$

By the law of Lambert-Beer, ${}^{I_t}/_{I_0} = e^{-abc}$, we obtain:

$$I_f = Y_f I_0(1 - e^{-abc}) \tag{18}$$

We study the function e^{-abc} by Taylor, we have

$$e^{-abc} = 1 - abc + (abc)^2/2! - (abc)^3/3! + \cdots \tag{19}$$

While abc is close 0, we have $e^{-abc} = 1 - abc$, taking to Eq.(3),

$$I_f = Y_f I_0 abc \tag{20}$$

Using the molar absorption coefficient ε take place a, we obtain:

$$I_f = 2.303 Y_f I_0 \varepsilon bc \tag{21}$$

We may find that, for some dilute fluorescent material solution, the fluorescence intensity is proportional to the concentration of the substance, while the frequency and intensity of laser is at certain.

4.1.3. Fluorescenc Resonance Energy Transfer Theory

Fluorescence (or Forster) resonance energy transfer is a process that involves nonradiative energy transfer from a photoexcited donor molecule, after absorption of a higher energy photon, to an acceptor molecule of a different species (brought in close proximity), which may relax to tis ground state by emitting a lower energy photon. This process results from dipole-dipole interactions and is thus strongly dependent on the center-to-center separation distance; it requires a nonzero integral of the spectral overlap between donor emission and acceptor absorption. Careful selection of an appropriate donor-acceptor pair ensures high transfer efficiency and provides two measurable parameters: a quenched donor photoemission and enhanced acceptor fluorescence. While good spectral overlap is paramount to high transfer efficiency, the donor and acceptor photoluminescence (PL) signals must also be well resolved to extract accurate experimental information for the system under investigation, e.g., a binding event due to a specific interaction in a biological receptor-ligand system or a change in protein conformation due to interactions with a target molecule. In addition, the excitation line should be chosen to coincide with the minimum of the acceptor absorption spectrum in order to reduce contributions resulting from direct excitation of acceptor.

The rate of energy transfer for an isolated single D-A pair separated by a distance r can be expressed using the Forster formalism as

$$k_{D-A} = \frac{B \times Q_D I}{\tau_D r^6} = \left(\frac{1}{\tau_D}\right) \times \left(\frac{R_0}{r}\right)^6 \tag{22}$$

where Q_D is the quantum yield of the donor and τ_D is the excited-state lifetime of the donor; the constant B is a function of the refractive index of the medium n_D, Avogadro's number N_A, and a parameter, k_p, that depends on the relative orientation of the donor and acceptor dipoles:

$$\mathrm{B} = \frac{[9000 \times (ln10)]\kappa_p^2}{128\pi^5 n_D^4 N_A} \tag{23}$$

where κ_p^2 = 2/3 for randomly oriented dipoles and varies between 0 and 4 for the cases of orthogonal and parallel dipoles, respectively. The overlap integral, I, defined as

$$\mathrm{I} = \int_0^{\infty} PL_{D-corr}(\lambda)\varepsilon_A(\lambda)\lambda^4 d\lambda, \tag{24}$$

is a quantitative measure of the donor-acceptor spectral overlap over all wavelengths λ; it is a function of the normalized donor emission spectrum (dimensionless property), PL_{D-corr}, and the acceptor absorption spectrum (expressed as an extinction coefficient), ε_A. In Eq.(22), we introduced the Forster radius (or distance), R_0, which is defined as

$$R_0 = (BQ_D I)^{1/6} = \left(\frac{9000(\ln 10)\kappa_p^2 Q_D}{N_A 128\pi^5 n_D^4}\right)^{1/6} \tag{25}$$

The FRET efficiency, E, defined as

$$\mathrm{E} = \frac{k_{D-A}}{k_{D-A}+\tau_D^{-1}} = \frac{R_0^6}{R_0^6+r^6} \tag{26}$$

accounts for the fraction of excitons that are transferred from the donor to the acceptor nonradiatively. For FRET within the quasi-monodispersed QD assembly, where these QDs act as donors, the FRET efficiency can be estimated as

$$\mathrm{E} = 1 - \frac{\tau_{DA}}{\tau_D}, \tag{27}$$

where τ_D and τ_{DA} are the lifetime of the donor in the absence and presence of the acceptor. The Forster theory treats the donor and acceptor as points in the interaction space, whereas the nanocrystals have finite size and are relatively large compared to the dye molecules.

4.2. Results and Discussion

4.2.1. The Analysis of the Steady-State Experiment

We use the steady-state fluorescence technique for studying three samples-deionized water solutions of CuPc, CdTe and CuPc-CdTe at room temperature. In this chapter, all of the steady-state spectras were measured at the same condition, and the solvent played unimportant role in the steady-state properties.

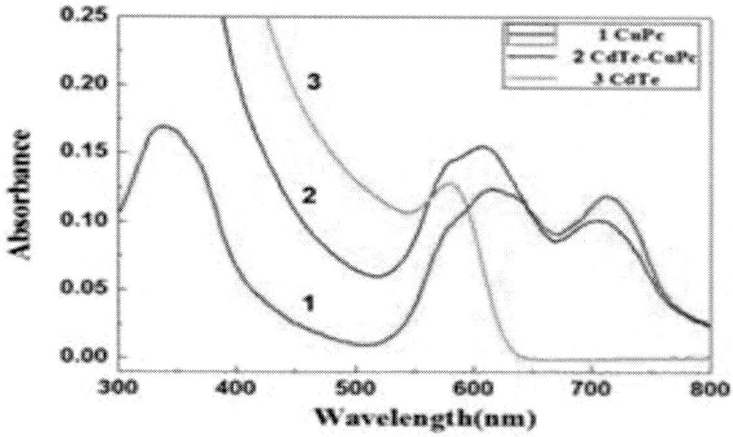

Figure 12. Steady-state absorption spectra for three samples: deionized water solutions of (1)CuPc, (2). CdTe-CuPc and (3)CdTe.

(1) The steady-state absorption spectra for three samples-deionized water solutions of CuPc, CdTe and CuPc-CdTe at room temperature are shown in Figure 12. Figure 12 depicts that CuPc spectrum consists of two strong bands(Q and Soret bands) at about 600-750 and 300-400nm. CdTe QDs spectrum has a broad peak with the lowest energy peak at 580nm. CdTe QD-CuPc spectrum has two broad peaks, with the maxima located at 608 and 702nm. If we use a 780nm femtosecond laser pulse to excite the CdTe QD-

CuPc solution, one photon energy is not enough to excite the electron on ground state to the excited state for energy transfer. Therefore, there are TPA processes when the QD-CuPc system is excited by a 780nm femtosecond laser pulse.

(2) The steady-state absorption spectra for samples-deionized water solutions of CuPc and fluorescence spectra CuPc-CdTe at room temperature are shown in Figure 13.

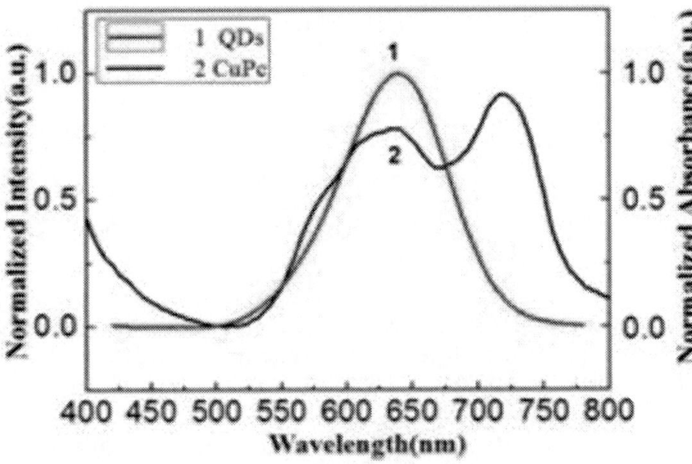

Figure 13. The emission spectrum for CdTe solution (the curve 1) and the absorption spectrum for CuPc solution (the curve 2).

Figure 13 shows that the luminescent maximum of the CdTe QDs solution is located ate 650nm, matching the Q-band absorption of CuPc. Thus, the conditions are met for energy transfer through the FRET mechanism.

(3) Next, we study the effect of the concentration on fluorescence spectra. Figure 14 is the steady-state fluorescence spectra of CdTe QDs, CdTe-CuPc and CuPc samples. Curve 1 is the fluorescence spectra of CuPc, the main emission peak is located at 825 nm, as well as a small emission peak located at 767 nm. Curve 5 is the fluorescence spectra of CdTe QDs, the emission peak is located at 650 nm. Curve 2 to Curve 4 is the fluorescence spectra of three CdTe-CuPc samples with different CuPc concentrations. It can be seen from Figure 15 that the fluorescence emission intensity of CuPc increases with an increase of CuPc concentration, while the fluorescence emission intensity of CdTe QDs decreases when the CuPc concentration increases. The

enhancement of CuPc fluorescence emission in solutions with CdTe QDs is strong evidence of energy transfer from QDs to Pc molecules.

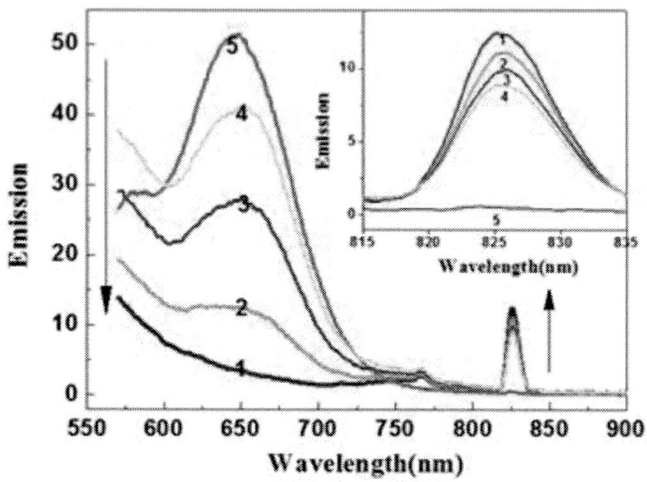

Figure 14. The steady-state fluorescence spectra with (1) CuPc(1.5 × 10-4mol/L) and three CdTe-CuPc samples with different CuPc concentrations (2) 1.0 × 10-4mol/L, (3) 7.5 × 10-5mol/L, (4) 1.5 × 10-5mol/L and (5)CdTe QDs.

4.2.2. The Analysis of the Time-Resolved Fluorescence Spectra

(1) The effect of concentration to FRET efficiency

We study the contribution of concentrations of copper phthalocyanine solution to the FRET efficiency. In this chapter, all of the time-resolved fluorescence spectras were measured at the same condition, and the solvent played unimportant role in the time-resolved fluorescence properties.

Five samples with fixed volume ratios of CdTe QD and CuPc solutions are prepared, where the concentration of CuPc is changed and that of CdTe QDs is fixed. Two-photon excited fluorescence decay curves of CdTe and the four CdTe–CuPc samples are shown in Figure 15. The excitation wavelength was 780nm, excitation laser power was 500mW, and monitoring wavelength was 650nm. The decay-component spectra of experimental data are fitted by the monoexponential model. We then obtain the value of the fluorescence lifetime τ for each curve. Figure 15 show that the fluorescence lifetime of CdTe–CuPc is less than that of CdTe under the same excitation laser power, indicating that there is energy transfer from CdTe (donor) to CuPc (acceptor) through the FRET mechanism. Figure 15 also show that the fluorescence lifetime decreases with an increase in CuPc concentration when the CuPc concentration is lower than 1.5×10^{-4} mol/L.

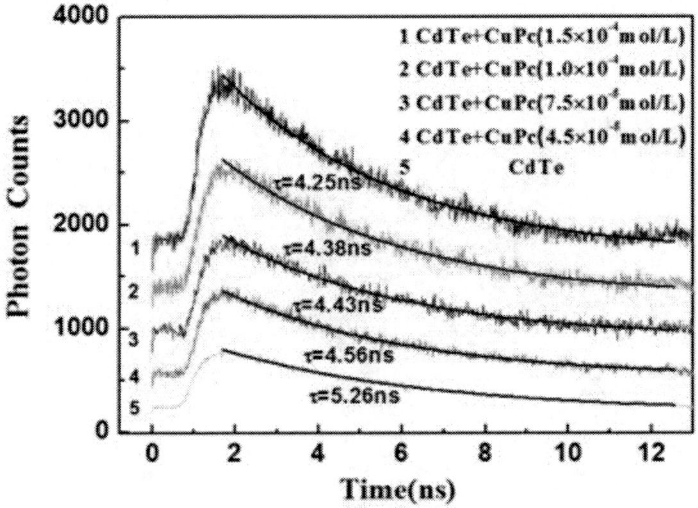

Figure 15. Two-photon excited fluorescence decay curves of CdTe and four CdTe-CuPc samples with different CuPc concentrations. The solid lines are the exponential fits. Excitation wavelength was 780 nm, excitation laser power was 500 mW, and monitoring wavelength was 650 nm.

The nonradiative energy transfer efficiencies of the four CdTe–CuPc samples can thus be calculated using equation (27). Table 2 shows the FRET efficiency obtained from the fitting curve shown in Figure 15.

Table 2. The FRET efficiency of different concentration of CuPc-CdTe mixed system

Concentration (10^{-5}mol/L)	15	10	7.5	4.5
E (%)	19.2	16.7	15.8	13.3

The table clearly shows that with the increasing of CuPc concentration in CdTe QDs-CuPc mixed system, the fluorescence resonance energy transfer efficiency is also growing. It is mainly due to the increasing effective molecular while the concentration of solution is growing. The optimal concentration of CuPc in CdTe–CuPc mixed system is 5×10^{-4}mol/L.

(2) The effect of laser power to FRET efficiency

Next, we study the contribution of laser power to the FRET efficiency. In this chapter, all of the time-resolved fluorescence spectras were measured at the same condition, and the solvent played unimportant role in the time-resolved fluorescence properties.

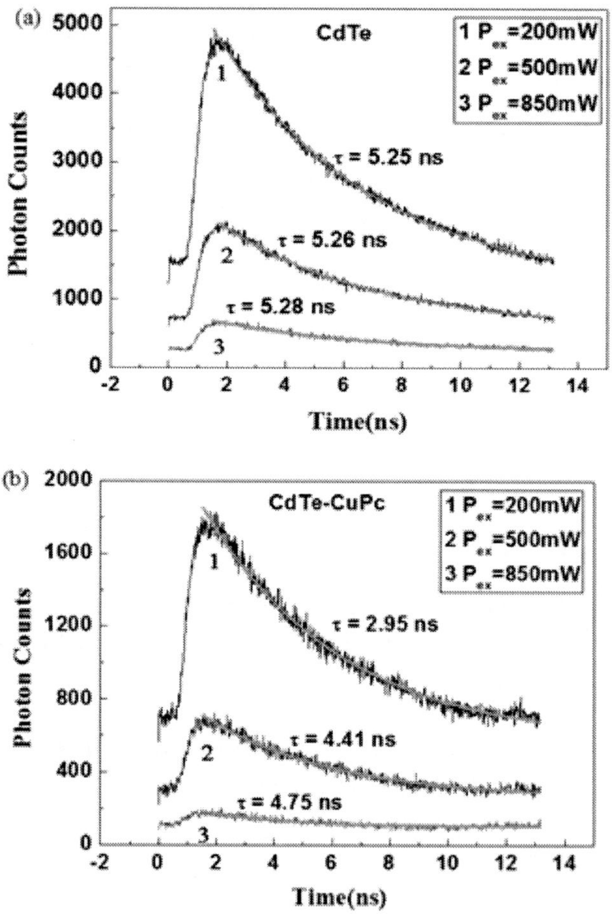

Figure 16. Two-photon excited fluorescence decay curves of (a) CdTe QDs and (b) CdTe QDs-CuPc with three excitation laser powers. The solid lines are the exponential fits. Excitation wavelength was 780 nm, and monitoring wavelength was 650 nm.

Here, we study the effect of the excitation laser power on the lifetime and nonradiative energy transfer efficiency. The excitation laser power is varied from 0 to 850mW. Figure 16 shows the two-photon excited fluorescence decay curves of CdTe and CdTe-CuPc for three excitation laser powers. The decay component spectra of experimental data are fitted by the monoexponential model. It can be seen from Figure 16 that the lifetime of CdTe-CuPc is shorter than that of CdTe under the same excitation laser power. This indicates that there is energy transfer from CdTe to CuPc.

It can be obtain form Figure 16 that the lifetime of CdTe is 5.25ns and the lifetime of CdTe in CdTe-CuPc complex system is 2.95ns when the excitation power is 200mW. The lifetime of CdTe is 5.26ns and the lifetime of CdTe in CdTe-CuPc complex system is 4.41ns when the excitation power is 500mW. The lifetime of CdTe is 5.28ns and the lifetime of CdTe in CdTe-CuPc complex system is 4.75ns when the excitation power is 850mW. It clearly shows that the fluorescence lifetime of CdTe QDs-CuPc lengthens with an increase in excitation laser power.

We obtain the FRET efficiency while the excitation laser power is different by using Eq.(27):

When the excitation laser power is 200mW, the FRET efficiency is

$$E = 1 - \frac{\tau_{DA}}{\tau_D} = 1 - \frac{2.95}{5.25} = 43.8\%.$$

When the excitation laser power is 500mW, the FRET efficiency is

$$E = 1 - \frac{\tau_{DA}}{\tau_D} = 1 - \frac{4.41}{5.26} = 16.2\%.$$

When the excitation laser power is 850mW, the FRET efficiency is

$$E = 1 - \frac{\tau_{DA}}{\tau_D} = 1 - \frac{4.75}{5.28} = 10.0\%.$$

It is found that the excitation threshold is about 200mW. It is also evident that the nonradiative energy transfer efficiency E decreases with an increase in excitation laser power. The maximum energy transfer efficiency can reach 43.8% via two-photon excitation with laser power of 200 mW.

The efficiency of energy transfer from the CdTe to phthalocyanine can be evaluated through Forster analysis:

$$E = \frac{nk_{D-A}}{nk_{D-A}+\tau_D^{-1}} = \frac{nR_0^6}{nR_0^6+r^6} \tag{28}$$

where k_{D-A} is the rate of energy transfer, r is the distance between the donor and acceptor, R_0 is the critical transfer distance or the distance at which half of the donor molecules decay by energy transfer, and n is the number of acceptor molecules per donor.

We calculate the efficiency of FRET between a single donor-acceptor pair because the sample solution is a non-conjugated system (i.e., taking n as 1):

$$E = \frac{R_0^6}{R_0^6+r^6} = \frac{1}{1+\left(\frac{r}{R_0}\right)^6} \tag{29}$$

Because the sample solution is a non-conjugated system of CuPc and CdTe QDs, the donor-acceptor distance r will slightly change according to the thermal motion of the molecules. If we use experimental values of E of 43.8%, 16.2% and 10.0% to calculate the donor-acceptor distance r using eq.(29), we obtain values of r of 6.8, 8.4 and 9.2nm when the excitation laser power is 200mW, 500mW, and 850mW, respectively. One can see that the donor-acceptor distance r increases with the excitation laser power. This is because the thermal motion of molecules increases as the excitation laser power increases. In other words, the decrease in energy transfer E is due to the larger donor and acceptor distance r at higher excitation laser power, which in turn is due to the thermal effect.

Furthermore, a tentative model of energy transfer between QDs and phthalocyanine with two-photon excitation is proposed to explain the above experimental phenomenon, which is described by Figure 17. The left part in Figure 17 shows the level structure of QDs: the conduction band minimum (CBM), electron surface states (ESSs), hole surface states (HSSs) and valance band maximum (VBM). The right part in Figure 17 shows the energy level structure of copper phthalocyanines: the ground state S_0, the first, second and the upper excited states S_1, S_2 and S_n. When using the 780-nm laser pulse to excite the CdTe QDs-copper phthalocyanines system, the QDs may absorb two photons (TPA) simultaneously, promoting an electron from the VBM to CBM. Some electrons in the ESSs will then return to the CBM or HSSs, emitting fluorescence. Because of the electron surface states of QDs and the first excited state of phthalocyanine are very close, some of electrons in the

ESSs will jump to the lowest vibrational level of the first excited state S_1 through nonradiative transition, which is called energy transfer. The thermal activation of molecules in the metastable level increases as the excitation laser power increases, and the electron in the higher metastable level will return to the defect states of CdTe QDs, this electron exchange is the inverse process of energy transfer, it leads to the energy transfer efficiency decrease. On the other hand, when using the 780-nm laser pulse to excite the CdTe QDs-copper phthalocyanine system, the copper phthalocyanine may also absorb two photons simultaneously, promoting an electron from S_0 to S_2 directly. When the excitation laser power is increased, more and more electrons in the S_2 state will absorb another photon and jump to the upper excited state S_n, which is called excited-state absorption (ESA) induced by TPA. This phenomenon results in the energy transfer efficiency E decreasing with increasing excitation laser power. As shown in Figure 12, copper phthalocyanine still has sizable absorption at the laser excitation wavelength of 780nm, and would be highly excited at higher laser power. Therefore, when the laser excitation power is increased, more and more electrons in the S_2 state will absorb one photon and jump to the upper excited state S_n, resulting in the electron transfer from the CdTe QDs electron surface states to CuPc becoming less efficient.

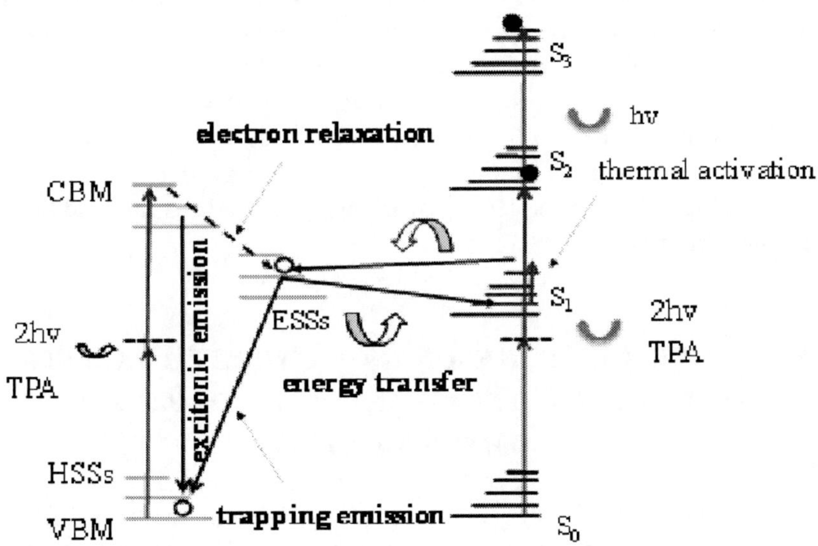

Figure 17. Schematic diagram of energy transfer between QDs and Pc under two-photon excitation. ○:Donor electron, ●:Acceptor electron.

4.3. Conclusion

In summary, the optical properties in the solution of copper phthalocyanine, CdTe QDs and CdTe QDs-copper phthalocyanine were observed by steady-state fluorescence spectra technique. It depicts that there are TPA processes when the QD-CuPc system is excited by a 780nm femtosecond laser pulse. And the fluorescence emission intensity of CuPc increases with an increase of CuPc concentration, while the fluorescence emission intensity of CdTe QDs decreases when the CuPc concentration increases. The enhancement of CuPc fluorescence emission in solutions with CdTe QDs is strong evidence of energy transfer from QDs to copper phthalocyanine molecules.

The photoinduced process in CdTe QDs- CuPc non-conjugated system via two-photon excitation was studied employing the time-correlated single-photon counting technique with femtosecond laser pulses. Efficient FRET in the CdTe QDs- CuPc system with two-photon excitation is observed. It is demonstrated that the fluorescence lifetime of CdTe–CuPc is less than that of CdTe under the same excitation laser power, indicating that there is energy transfer from CdTe (donor) to CuPc (acceptor) through the FRET mechanism. Besides, the fluorescence resonance energy transfer efficiency is also growing, with the increasing of CuPc concentration in CdTe QDs-CuPc mixed system. It is mainly due to the increasing effective molecular while the concentration of solution is growing. The optimal concentration of CuPc in CdTe–CuPc mixed system is 5×10^{-4}mol/L, the FRET efficiency is 43.8%. Furthermore, the decrease in energy transfer E is due to the larger donor and acceptor distance r at higher excitation laser power, owing to the ESA induced by TPA and the thermal effect.

5. Fluorescence Resonance Energy Transfer Experiments of CdTe Quantum Dots-Rhodamine B Composite System

Fluorescent dyes of Rhodamine is widely used in medicine, biology, environmental chemistry, which is due to the special structure and corresponding fluorescence properties, such as the excellent light-stability, insensitivity to pH, wide range of wavelength and the higher fluorescence quantum yield, it becomes the most commonly used dyes in field of chemical and biological technology [56-57].

In this part, a two-photon FRET aqueous system, using CdTe QDs as the donors and Rhodamine B (RhB) as the acceptor, was investigated and analyzed. Steady-state photoluminescence (PL), time-resolved two-photon PL and fluorescence lifetime measurements were performed to enable analysis of the kinetics of the two-photon excited luminescence. Furthermore, the relationship between the extent of the spectral overlap and the FRET efficiency was also discussed.

5.1. Optical Experiments of CdTe QDs-RhB Composite System

5.1.1. Experimental Materials

CdTe QDs were prepared from Cd^{2+} and NaHTe in the presence of thioglycolic acid in aqueous solution by electrostatic reaction. RhB, the acceptor, was purchased from Aladdin Reagent Co., Ltd and was prepared as an aqueous solution. The concentrations (unit: mol/L) of the two solutions are of the 10^{-6} ~10^{-5} order of magnitude to ensure that there is little reabsorption during PL measurement. All experiments were performed with ultrasonic oscillator after preparing the mixture solutions.

The steady state photoluminescence (SPPL) was studied with a JASCO FP-6500 Fluorescence Spectrometer with 1mm quartz cuvettes at room temperature. The fluorescence spectrometer uses a xenon lamp as an excitation source and thus the fluorescence spectroscopy result can be considered to be measured under one-photon excitation. To avoid any interference caused by the absorption of the RhB acceptor, the excitation wavelength was kept at 400nm which is away from RhB absorption peak. The slot set as 3mm-3mm was considered to be suitable. The absorption measurements of solutions were undertaken using a PerkinElmer Lambda 35 UV-Vis spectrophotometer.

5.1.2. Experimental Setup

To analyze the kinetics of two-photon excited luminescence, time resolved two-photon photoluminescence and fluorescence lifetime measurements are performed by the time-resolved fluorescence test system shown in Figure18. The samples are excited at 800nm wavelength with femtosecond laser pulses, which is produced by Titanium sapphire femtosecond laser whose pulse widths of 130fs, repetition frequency of 76MHz, and the power is fixed at 500mw. The fluorescence spectroscopy is measured by fluorescence spectrometer and the fluorescence decay curves are recorded by single photon counter. The spectra are recorded by EM-CCD (Andor DU970P) through monochromator (Andor SR500). The samples are carried by 10mm cuvette and in order to avoid the interference of the excitation source, the cuvette installed at 45 degree angle so that the spectrometer can collect the reflection fluorescence of the sample. The filters are also used for the same purpose.

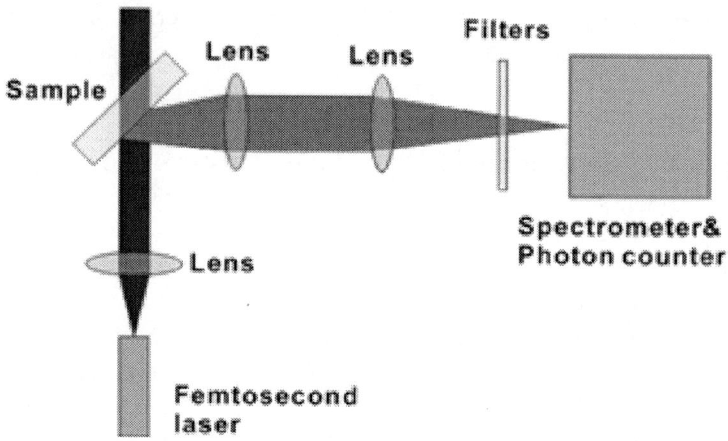

Figure 18. The optical path of time-resolved fluorescence test system.

5.2. Results and Discussion

5.2.1. Absorption Spectra and Steady-State Photoluminescence

The normalized absorption spectra of three CdTe QDs samples (i.e., QDs1, QDs2 and QDs 3) are shown in Figure 19(a). The absorption peak of QDs1, QDs2 and QDs 3 was located at 480nm, 549nm and 580nm, respectively. The mean grain size of the CdTe QDs is calculated to be 1.69nm, 3.23nm and 3.50nm, by using equation

$$D = (9.8127 \times 10^{-7})\lambda^3 - (1.7147 \times 10^{-3})\lambda^2 + (1.0064)\lambda - 194.84 \tag{30}$$

where D is the particle size, is the wavelength of the first excitonic absorption peak of the correspongding sample [58]. The normalized RhB absorption spectrum and the normalized SSPL of the QDs are given in Figure 20(b), it shows the absorption spectrum of RhB dye which is located at 553 nm and emission spectra of QDs1, QDs2 and QDs3, which are located at 522nm, 597nm and 650nm, respectively. The concentrations of different QDs are 5.64×10^{-6} mol/L. Spectral overlap between the donor emission spectrum and the acceptor absorption spectrum is essential for FRET, and the energy transfer efficiency is closely related to the extent of this spectral overlap. Figure 19(b) shows that the largest spectral overlap is found between the fluorescence spectrum of QDs1 and the RhB absorption spectrum, followed by that with the spectrum of QDs2, while QDs3 has the smallest spectral overlap with the RhB absorption spectrum.

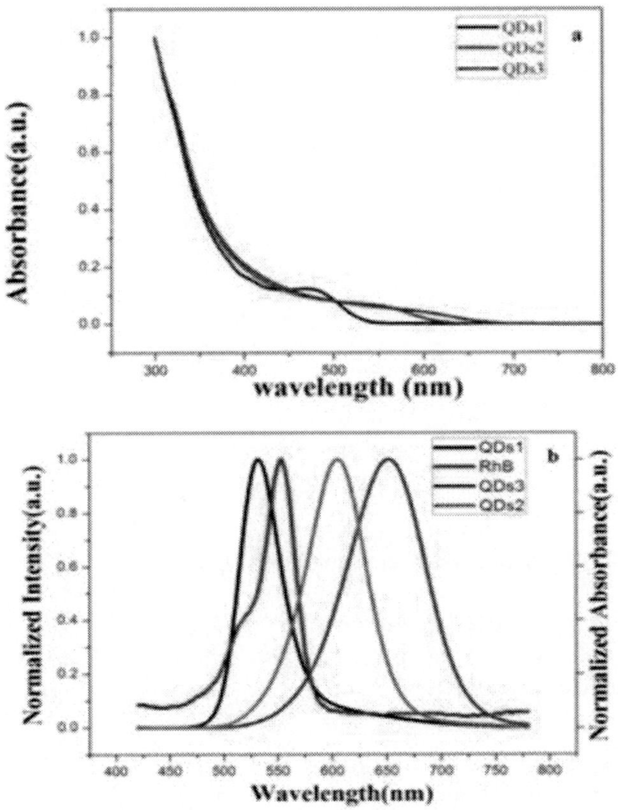

Figure 19. (a) Absorption spectra of three CdTe QDs and (b) Normalized RhB absorption spectrum and the PL spectra of the QDs, where the PL peaks of QDs1, QDs2 and QDs3 are 530 nm, 607 nm, and 650 nm, respectively. The RhB concentration is 2.82×10^{-5} mol/L and the QDs have concentrations of 5.64×10^{-6} mol/L.

The SSPL spectra of the samples with different acceptor to donor (A/D) concentration ratios are shown in Figure 20. In these mixture solutions, the concentration used for all QDs sizes is 2.821 × 10^{-6} mol/L) and RhB with different concentrations as follows: RhB1: 2.75 × 10^{-6}mol/L, RhB2: 7.50 × 10^{-6}mol/L, RhB3:1.41 × 10^{-5}mol/L, RhB4:2.25 × 10^{-5}mol/L, RhB5:3.00 × 10^{-5} mol/L. This Figure shows that the fluorescence intensity of the QDs decreases when the A/D concentration ratio increases, indicating significant fluorescence quenching of the donor, and enhancement of the acceptor.

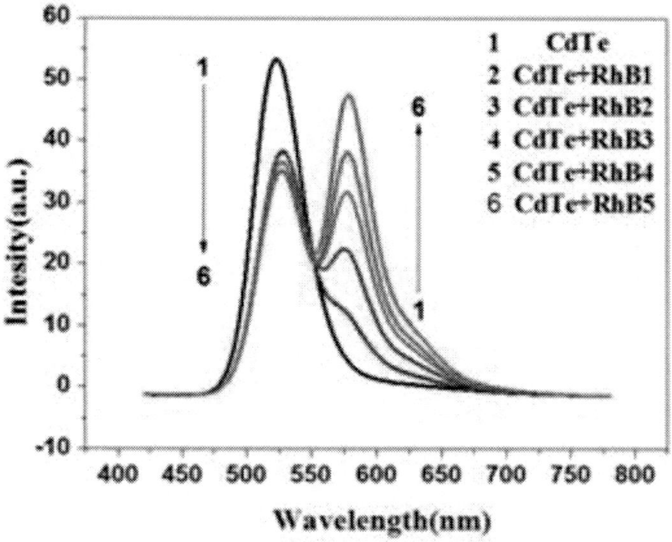

Figure 20. SSPL spectra of the mixture solutions of QDs1.

5.2.2. The Impact of RhB's Concentrations to QDs-RhB System

Figure 21 shows the two-photon fluorescence spectra of the QDs-RhB solutions. Because of the experimental error caused by sample replacement, the spectra have been normalized and plotted for the sake of the discussion. Because QDs1 has the largest spectral overlap of the three sizes of QDs, QDs1 is chosen to measure both the time-resolved photoluminescence and the fluorescence lifetime for various A/D concentration ratios, and the results are plotted in Figure 21, it indicates that the PL peak intensity of the QDs decreases until it almost disappears, whereas the A/D concentration increases. Therefore, the FRET between the QDs and the RhB in solution can be considered to be saturated at high A/D concentration ratio.

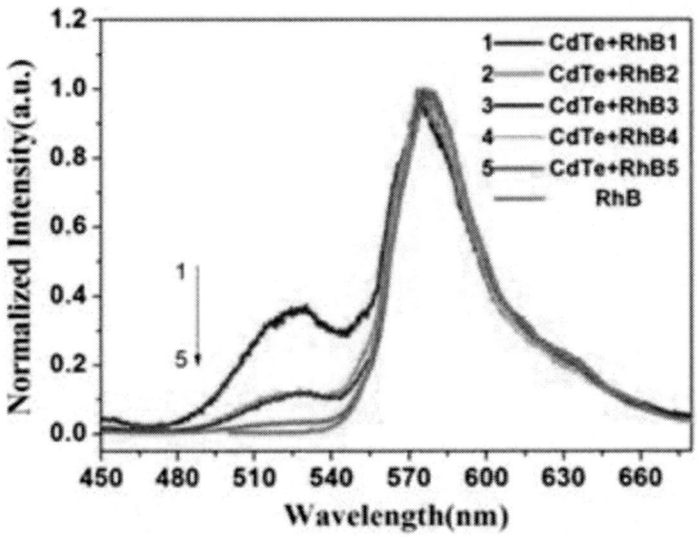

Figure 21. Two-photon fluorescence spectra of QDs1-RhB solutions with various A/D concentration ratios.

The fluorescence decay curves of the QDs1 and RhB mixed solutions with various A/D concentration ratios are measured using a single photon counter which selects the fluorescence peak wavelengths of RhB (575nm) and QDs (530nm) as the sampling points, with the integration time set at 900s. The measurements were performed in a darkroom at room temperature to reduce the measurement error caused by natural light. The decay curves were fitted using the following exponential decay curve

$$y = A\exp(-x/t)+y_0 \tag{31}$$

where t is the two-photon fluorescence lifetime. Figure 22(a) shows each decay curve with its corresponding exponential fitting curve. Figure 22(b) shows that the A/D concentration ratio becomes higher, the higher resulting photon count can be obtained. Therefore, it can be concluded that a higher A/D concentration ratio produces a stronger RhB fluorescence intensity. Table 3 shows the lifetimes obtained from the fitting curve shown in Figure 22(b). The lifetime of RhB is shown to be prolonged with increasing A/D concentration ratio, while the lifetimes of the QDs are reduced.

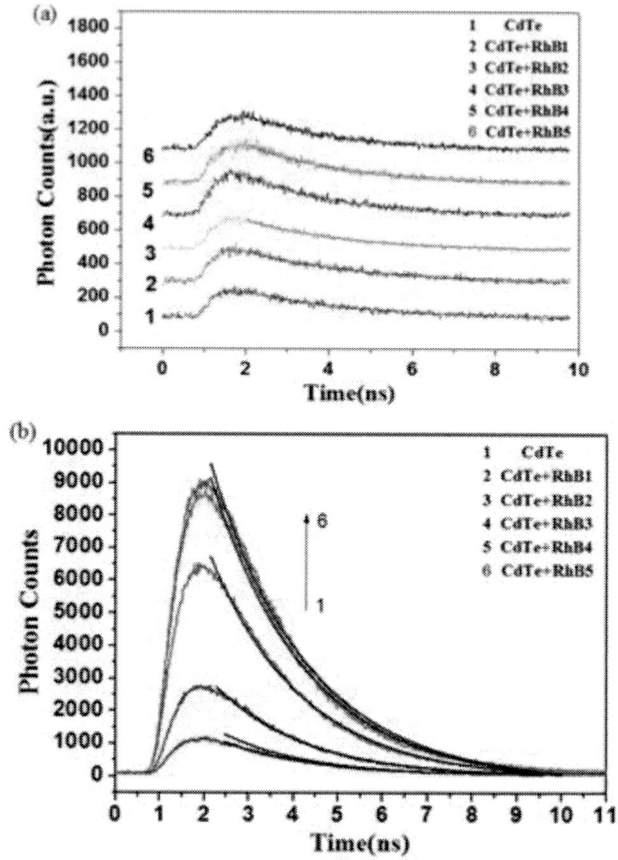

Figure 23. Fluorescence decay curves of QDs1-RhB solutions with various A/D concentration ratios and their corresponding fitting curves at (a) a wavelength of 530 nm (the fluorescence peak wavelength of QDs1) and (b) a wavelength of 575 nm (the RhB fluorescence peak wavelength).

5.2.3. The Impact of QDs' Size to QDs-RhB System

The two-photon fluorescence spectra of the mixture solutions of the different QDs and RhB are shown in Figure 23. Because the fluorescence peaks of RhB and QDs2 are too close together, peak-fitting analysis is performed in Figure 23(b) for ease of comparison. It shows the sequence of degree of fluorescence quenching ranked in order of QDs1, QDs2 and QDs3. It can be concluded that the fluorescence is quenched with increasing spectral overlap.

Table 3. The fitted fluorescence lifetime of QDs1 and RhB mixed solutions with various A/D ratios

Sample	QDs lifetime (ns)	RhB lifetime(ns)
RhB		1.92
CdTe	2.99	
CdTe+RhB1	2.21	1.95
CdTe+RhB2	1.86	1.99
CdTe+RhB3	1.84	2.06
CdTe+RhB4	1.81	2.13
CdTe+RhB5	1.81	2.18

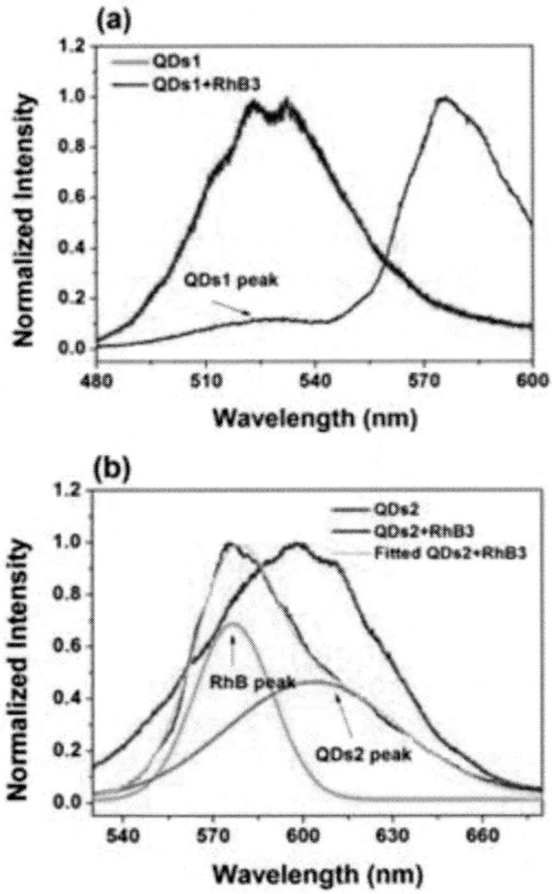

Figure 23. (Continued)

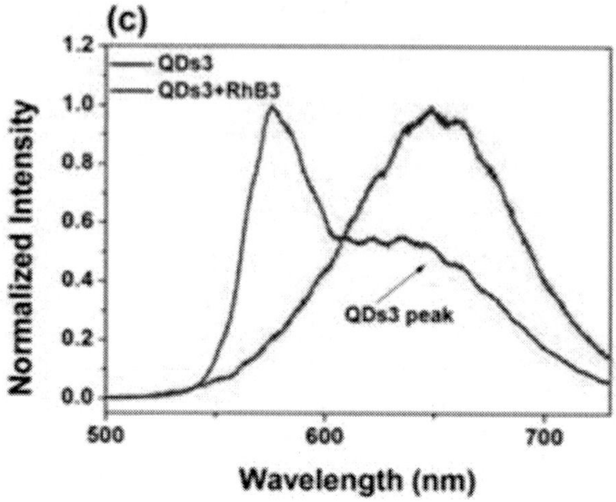

Figure 23. Two-photon fluorescence spectra of (a) QDs1+RhB3, (b) QDs2+RhB3 and (c) QDs3+RhB3. Note that the curve for "QDs2+RhB3" has been peak-fitted for clear visibility.

Table 4. Fitted fluorescence lifetime of the QDs-RhB solutions with different QD emission wavelengths

Sample	QDs lifetime (ns)	RhB lifetime(ns)
RhB3		1.83
QDs1	2.99	
QDs1+RhB3	1.86	2.06
QDs2	3.07	
QDs2+RhB3	2.10	2.12
QDs3	4.26	
QDs3+RhB3	2.94	2.20

At the same time, we measure the fluorescence decay curves of the QDs-RhB solutions with the different QD emission wavelengths, the peak emission wavelength of 575 nm for RhB, and the peak wavelengths of 530 nm, 607 nm and 650 nm for the QDs are chosen as sample points. With the same experimental conditions, Figure 24 shows the measurement results. Figure 24(a) indicates that the photon counts for RhB in the samples increase with increasing spectral overlap. This result suggests that the RhB fluorescence intensity varies directly with the degree of spectral overlap. Abnormally high photon counts for the sample "QDs2+RhB3" are shown in Figure 24(c). This phenomenon may be caused by the closeness of the sample point to the fluorescence peak of RhB which indicates that there may be some interference.

The fluorescence lifetimes are calculated by fitting the exponential decay curve of eq. (31) and the results are shown in Table 4. The prolonged RhB lifetime and the reduced QDs lifetimes can also be found in Table 4.

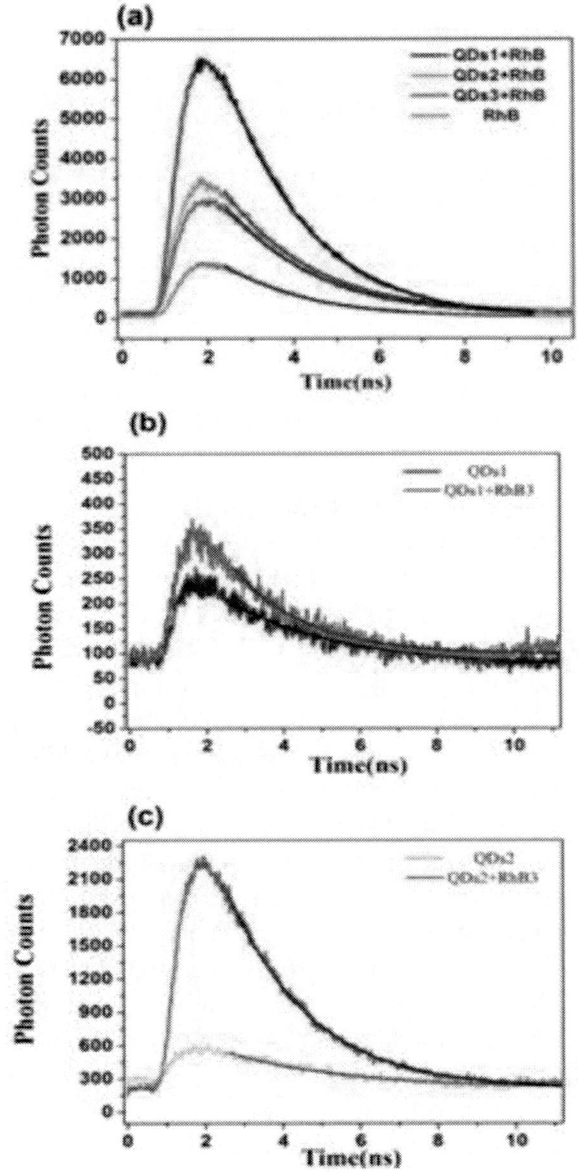

Figure 24. (Continued)

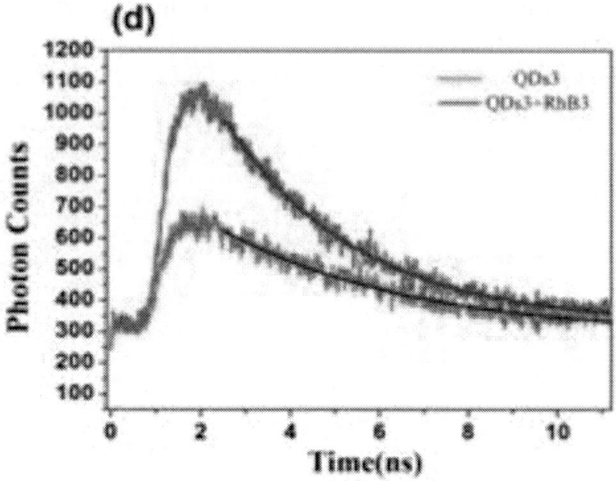

Figure 24. Fluorescence decay curves and their corresponding fitting curves at (a) a wavelength of 575 nm (the fluorescence peak wavelength of RhB), (b) 530 nm, (c) 607 nm and (d) 650 nm (the fluorescence peak wavelengths of the three sizes of QDs).

5.2.4. The Analysis of FRET Properties of QDs-RhB System

Forster resonance energy transfer is widely used for the detection of interactions between a directly excited emitter of fluorescence (donor) and a molecule of an energy acceptor. The efficiency of FRET depends on several factors including: distance between the donor and the acceptor, spectral overlap of the donor emission and the acceptor absorption spectra and relative orientation of the donor emission and the acceptor absorption transition dipole moments. In other words, the donor in the excited states while the distance of the donor-acceptor pair is appropriate, what's more, the energy level difference of the states of the donor-acceptor with the first excited electronic state are fitting, can transfer energy to acceptor through the dipole-dipole interaction, and stimulates the acceptor. It is the process of FRET. Figure 25 is the schematic diagram of forster resonance energy transfer. The left part in Figure 25 shows the level structure of QDs: the ground state S_0 and the first excited states S_1. The right part in Figure 25 shows the energy level structure of Rhodamine B: the ground state S_0 and the first excited states S_1. The FRET process occurs to effects: Weaken of the fluorescence of donor while enhancement of the fluorescence of acceptor and the decreasing of the lifetime of donor while increasing of the lifetime of acceptor.

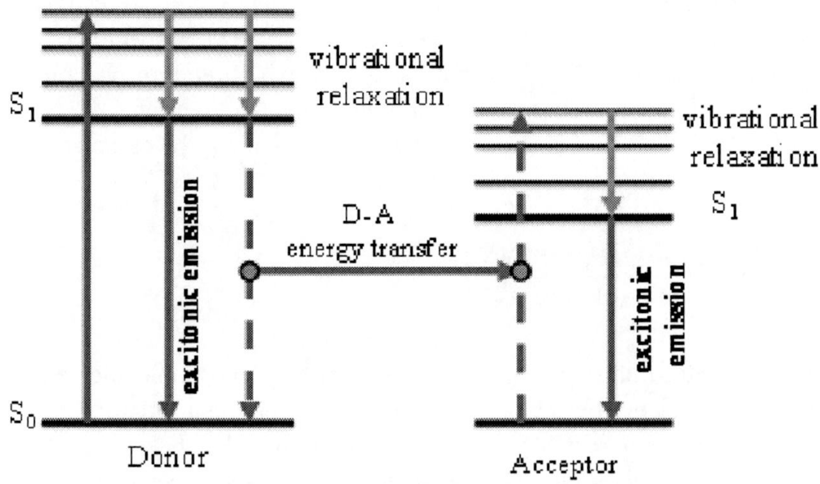

Figure 25. Schematic diagram of Forster resonance energy transfer.

Fluorescence lifetime is important, as it determines the time available for the fluorophore to interact with or diffuse in its environment. The lifetime of the excited state is defined by the average time the molecule spends in the excited state prior to return to the ground state. In the FRET process, because of the electron surface states of QDs and the first excited state of RhB are very close, some of electrons in the S_1 will jump to the lowest vibrational level of the first excited state S_1 through nonradiative transition, which is called energy transfer, it leads to the decreasing of the lifetime of donor, owing to accelerating the process of QDs return to the ground state.

Based on analysis of the results of the two-photon TRPL measurements, a reference method is introduced to estimate the two-photon absorption cross-section, and the corresponding formula is presented as eq. (32)

$$\delta_2 = \delta_1 \frac{F_2}{F_1}\frac{\phi_1}{\phi_2}\frac{c_1}{c_2} \tag{32}$$

where the subscript 1 represents the reference sample and subscript 2 represents the measured sample. δ is the two-photon absorption cross-section and has unit of GM (where $1GM = 10^{-50}$ $cm^4 \cdot s$/photon). F is the two-photon photoluminescence intensity. ϕ is the quantum yield and c is the sample concentration. RhB is chosen as the reference sample because its two-photon absorption cross-section is approximately 210 GM. From eq. (32), the two-photon absorption cross-sections of QDs1, QDs2, and QDs3 are estimated to be 4.68×10^3 GM, 8.20×10^3 GM, and 1.45×10^4 GM, respectively. These results show that the CdTe QDs that have high two-photon absorption cross-sections are suitable as donor for two-photon FRET.

The FRET characteristics of the QD-RhB system can be calculated using eq. (33):

$$R_0^6 = \frac{9000\ln(10)\kappa^2 Q_D J(\lambda)}{128\pi^5 n^4 N_A} \tag{33}$$

where the Förster radius, R_0, means that half of the donor molecules decay by energy transfer while the other half decay by conventional radiative and non-radiative mechanisms at this distance. In other words, the energy transfer efficiency is 50% when the distance between donor and acceptor is R_0. κ is a factor that describes the relative spatial orientation of the transition dipoles of the donor and the acceptor. When considering the random orientation in solutions, κ2 is usually assumed to be equal to 2/3. QD is the quantum yield of the donor in the absence of the acceptor and can be given by a reference method. NA is Avogadro's number. n is the refractive index of the mediumand J(λ) is spectral overlap integral, which describes the degree of spectral overlap between the PL of the donor and the absorption of the acceptor, and can be calculated using eq.(34):

$$\mathrm{J}(\lambda) = \int_0^\infty F_D(\lambda)\varepsilon_A(\lambda)\lambda^4 \mathrm{d}\lambda \tag{34}$$

where $F_D(\lambda)$, as a dimensional number, is the normalized fluorescence spectrum. Note that the source of the data for the data for $F_D(\lambda)$ is the two-photon results obtained in this experiment. $\varepsilon_A(\lambda)$ is the molar extinction coefficient of the acceptor and has units of $M^{-1} \cdot cm^{-1}$. $\varepsilon_A(\lambda)$ can be determined by the Lambert-Beer law. λ is the wavelength which is expressed in units of nm [59]. The values of $J(\lambda)$ and Q_D are presented in Table 5. Note that RhB is selected as the standard sample. The Q_D of RhB in an aqueous solution is 0.31.

Equation (33) can be simplified into eq. (35) in the form:

$$R_0^6 = 8.79 \times 10^{-5} \kappa^2 n^{-4} Q_D J(\lambda), \tag{35}$$

Table 5. Quantum yield Q_D of the QDs with different emitting wavelength and their corresponding spectral overlap integrals $J(\lambda)$

Sample	QD (%)	J(λ)
QDs1	11.0	7.25 × 1016
QDs2	4.70	5.00 × 1016
QDs3	10.8	1.33 × 1016

where R_0 is expressed in angstroms (Å). It is noticeable that eq. (35) is only used when λ is expressed in units of nm and $\varepsilon_A(\lambda)$ is expressed in $M^{-1}\cdot cm^{-1}$. Equation 35 indicates that R_0^6 is proportional to both J(λ) and Q_D, which means that higher a degree of spectral overlap and a higher donor quantum yield make the Förster radius larger, i.e., a wider energy transfer range can lead to higher FRET efficiency. Using eq. (35), the calculated results for the Förster radius R_0 are 7.40 nm, 6.03 nm, and 5.56 nm for QDs1, QDs2 and QDs3, respectively. Based on this observation, QDs1 has the widest Förster interaction range and is the best match to the acceptor among the three QD sizes.

The FRET efficiency E can be calculated using eq. (36) as

$$E = 1 - \frac{\tau_{DA}}{\tau_D}, \tag{36}$$

where and are the fluorescence lifetime of the donor with and without the presence of the acceptor, respectively. Based on the values of R_0 from eq. (35) and E from eq. (36), the donor-to-acceptor distance r can be calculated using the following equation:

$$E = \frac{R_0^6}{R_0^6 + r^6}. \tag{37}$$

The FRET efficiency is calculated for the QDs1 and RhB mixed solutions with various A/D concentration ratiosusing Eq. (36), and the results are shown in Figure 26. It is shown that as the A/D ratio increases, the lifetime of the donor (QDs) dropps accordingly, while the FRET efficiency increases logarithmically and can be as high as 40%. The acceptor lifetime increases linearly. It can be concluded that the FRET gradually saturates with increasing A/D ratio.

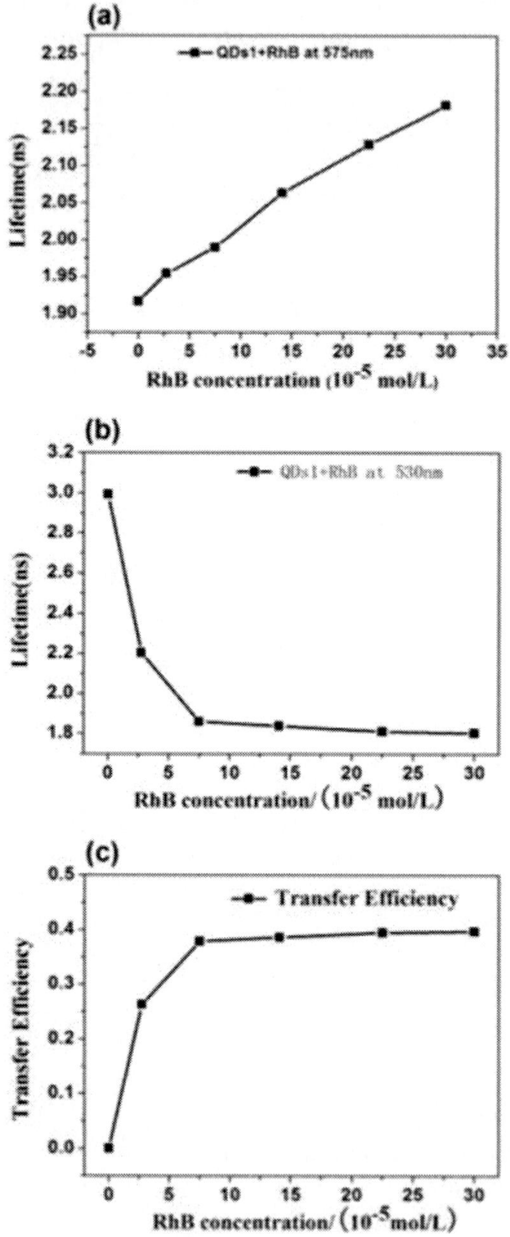

Figure 26. In the QDs1-RhB mixed solution which varying the A/D concentration ratio, (a) the fluorescence lifetime of RhB, (b) the fluorescence lifetime of QDs1 and (c) the FRET efficiency of QDs1-RhB.

Table 6 shows the results for FRET efficiency and the donor-to-acceptor distance for the mix solutions of QDs with different emission wavelengths and RhB and shows that there is a proportional relationship between the FRET efficiency and the degree of spectral overlap. The experimental FRET efficiency results agree approximately with the theoretically predicted values. While the distance r of the QDs1+RhB3 sample is the longest among the three samples, it is numerically closest to the corresponding R_0 and thus the QDs1+RhB3 sample can reach the highest FRET efficiency.

Table 6. FRET efficiency values of QD-RhB solutions

Sample	FRET efficiency E(%)	Förster radius R_0(nm)	donor-to-acceptor distance r(nm)
QDs1+RhB3	38.0	7.40	8.03
QDs2+RhB3	31.6	6.03	6.85
QDs3+RhB3	30.8	5.56	6.36

5.3. Conclusion

We achieved FRET in QDs-RhB aqueous solutions using two-photon excitation and calculated the Förster radii of the samples to be 7.40 nm, 6.03 nm, and 5.56 nm for QDs1, QDs2 and QDs3, respectively. The experimental results and the theoretical analysis indicated that the FRET efficiency increase with an increase of the spectral overlap, which is caused by an increase in the Förster radius R_0. In addition, we showed that the FRET efficiency increases with increasing acceptor/donor concentration ratio. The transfer efficiency can be as high as 40%. This study shows that there is a bright future for these materials in bio-probing, bio-imaging, bio-labelling and other biological and optoelectronic applications.

6. Summary and Prospect

With the development of nano-technology, quantum dot (QD) has been studied and paid more attention by scientists and other researchers. Compared with traditional organic fluorescent dyes, quantum dots have many special optical properties due to the wide fluorescence excitation spectrum, narrow and tunable emission spectrum, the higher fluorescence intensity and the excellent property of photobleaching, which have a wide range of applications in the field of chemical detection, biomedicine, electronics, etc. The traditional organic fluorescent dyes, such as Copper phthalocyanine and Rhodamine B, also have wide applications in the high-technology field due to their special properties of optics, electronics and magnetism. This article is mainly to study the CdTe quantum dots and organic fluorescent dyes composited system by using two-photon excited technique, focusing on the fluorescence resonance energy transfer of CdTe QDs-CuPc and CdTe QDs-RhB composite system from theory and experiment. The summary is below.

6.1. Research Contents and Conclusion

Based on reading, analyzing and summarizing the CdTe QDs-organic fluorescent dyes composited system and fluorescence resonance energy transfer theory in detail, the main contents and conclusions in this thesis are as follows:

(1) The properties and applications of the quantum dots were introduced, as well as their research development at home and abroad.

(2) Introducing the fluorescence resonance energy transfer theory. Illustrating the fluorescence excitation spectra and fluorescence absorption spectrum and introducing the principle of fluorescence resonance energy transfer and the research progress, which provides the theoretical basis for studying and constructing the CdTe QDs-organic fluorescent dyes composited system.

(3) The fluorescence resonance energy transfer (FRET) in CdTe quantum dots (QDs)-copper phthalocyanine (CuPc) is investigated by ultrafast time-resolved spectroscopy technique equipped with femtosecond laser (780nm,76MHz,130fs). The results show that the fluorescence lifetime of CdTe quantum dots decreases with the increase of CuPc concentration, and the energy transfer efficiency is found to increase with the increase of CuPc concentration. Moreover, the influence of the laser excitation power on the energy transfer efficiency is also studied. It is found that transfer efficiency decreases as excitation laser power increases, the physical mechanism is the thermal activation in the high power and the excited state transitions of high order induced by two-photon. The energy transfer efficiency can reach 43.8%, when the laser power is 200 mW, via two-photon excitation. This study

indicates that the CdTe QDs-CuPc composite system has high potential as the third generation of photosensitizers.

(4) The fluorescence resonance energy transfer (FRET) between CdTe quantum dots (QDs) with different sizes and Rhodamine B (RhB) in aqueous solution is investigated by using the time-resolved fluorescence test system under two-photon excitation. With the excitation of 800nm femtosecond laser, result shows that the FRET efficiency of the QDs-RhB system increases with the spectral overlap increase of the CdTe emission spectrum and the Rhodamine B absorption spectrum. Specifically, the fluorescence intensity of QDs decreases and the lifetime of QDs decreases while RhB shows the opposite tendency. By means of the Förster Theory of Energy Transfer, the spectral overlap integral $J(\lambda)$, Foster radius R_0 and the FRET efficiency E are calculated and The FRET characteristics of QD-RhB system is characterized. Theoretical analysis reveals that the physical source is the increase of the sample's Forster radius. Moreover, the relationship of the ratio of acceptor/donor concentration and the FRET efficiency is investigated experimentally. When the ratio of acceptor/donor concentration increases, the lifetime of QDs gets shorter, and the FRET efficiency of the QDs-RhB system becomes higher. The two-photon excited FRET efficiency can reach to 40.1% when the concentration of RhB is 3.0×10^{-5}mol/L. This study shows that there is a bright future for these materials in bio-probing, bio-imaging, bio-labelling and other biological and optoelectronic applications.

6.2. Feature and Innovation

(1) The research is based on the theory that two photon absorption (TPA) induced by femtosecond laser. The study of ultrafast dynamics of two-photon-excited QDs and organic dye energy transfer has not yet been reported while there is a lot of researching work about single-photon excitation at home and abroad.

(2) Two-photon excited energy transfer in a non-conjugated system of CdTe QD-copper phthalocyanine (CuPc) using time-resolved fluorescence spectra was first put forward. The effects of laser excitation intensity on the fluorescence lifetime and energy transfer efficiency were discussed. Furthermore, the physical mechanism of energy transfer in the QD-phthalocyanine system under two-photon excitation was also discussed, which supports theoretical direction for building the new type of photosensitizes.

(3) A two-photon fluorescence resonance energy transfer aqueous system, using CdTe QDs as the donors and Rhodamine B (RhB) as the acceptor, is investigated and analyzed. Steady-state photoluminescence (PL), time-resolved two-photon PL and fluorescence lifetime measurements were performed to enable analysis of the kinetics of the two-photon excited luminescence. Furthermore, the relationship between the extent of the spectral overlap and the fluorescence resonance energy transfer efficiency was also discussed, which supports theoretical direction for building the new type of biological fluorescence probe.

(4) Discussing the progress of fluorescence resonance energy transfer between QDs and organic dyes through femtosecond time-resolved ultrafast spectroscopy. Previous studies mainly employed steady-state absorption and fluorescence spectra technique for investigating the FRET between QDs-organic dyes, there is few report involving time-resolved dynamics. In the paper, photoinduced energy transfer in the QDs-organic dyes system via two-photon excitation is studied employing the time-correlated single-photon counting technique with femtosecond laser pulses, which reveals the physical mechanism of between the QDs and organic dyes deeply.

6.3. Prospect

(1) Fluorescence resonance energy transfer techniques is an effective mean for studying biological macromolecular structure as optical scale, it gradually combines with time-resolved fluorescence assay, fluorescence microscopy, etc. In future, It will play an important role in organelles, immunoassay and nucleic acid detection in the development of biological detection.

(2) In recent years, the light-induced energy transfer process between quantum dots – the organic dye has aroused the interest of many researchers, because QDs – organic dye composite system is the very promising third-generation photosensitizer and biological fluorescent probe, which has a vital significance in the biomedicine and treatment of medical tumor.

(3) With the improvement of laser technique, it provides the speedy development the new materials, new influences and applications through considering the FRET process between QDs and organic dyes. From the viewpoint of physics, with the emergence of femtosecond laser, a lot of novel physical and chemical phenomena start producing while the ultra-short and ultra-intense laser interaction with metal organic compounds, it is meaningful

to explore the deep physical mechanism and the application technology expands new field.

ACKNOWLEDGMENTS

This research was supported by the National Natural Science Foundation of China (Grant No. 11204222) and the Natural Science Foundation of Hubei Province in China (Grant No. 2013CFB316 and Grant No. 2014CFB793).

REFERENCES

[1] R P Feynman. There's plenty of room at the bottom. *Engineering and science*. 23(1960) 22-36.
[2] A. P. Alivisatos. Perspectives on the physical chemistry of semiconductor nanocrystals. *J. Phys. Chem. B*. 100(1996) 13226-13239.
[3] V. L. Colvin, A. P. Alivisatos, J. G. Tobin. Valence-band photoemission from a quantum-dot system. *Phys.Rev.Lett.* 66(1991) 2786.
[4] A. N. Goldstein, C. M. Echer, A. P. Alivisatos. Melting in semiconductor nanocrystals. *Science.* 256(1992) 1425.
[5] A. Henglein. Small-particle research: physicochemical properties of extremely small colloidal metal and semiconductor particles. *Chem.Rev.* 89(1989) 1861-1873.
[6] L. E. Brus. A simple model for the ionization potential electron affinity and aqueous redox potentials of small semiconductor crystallites. *J. Chem. Phys.* 79(1983) 5566-5571.
[7] R. J. Bandaranayake, G. W. Wen, J Y Lin. Structural phase-behavior in II-VI semiconductor nanoparticles. *Appl. Phys. Lett.* 67(1995) 831-833.
[8] G. L. Charron, T. Stuchinskaya, D. R. Edwards, et al.. Insights into the Mechanism of Quantum Dot-Sensitized Singlet Oxygen Production for Photodynamic Therapy. *J Phys Chem C*. 116(2012) 9334-9342.
[9] G. P. Mitchell, C. A. Mirkin, R. L. Letsinger. Programmed Assembly of DNA Functionalized Quantum Dots. *J. Am. Chem. Soc.* 121(1999) 8122-8123.
[10] Luo G, Long J, Zhang B, et al.. Quantum dots in cancer therapy. *Expert Opin.Drug Deliv.* 9(2012) 47-58.

[11] V. G. Praig, H Mcilwee, C. L. Schauer. Localized Surface Plasmon Resonance of Gold Nanoparticle-Modified Chitosan Films for Heavy-Metal Ions Sensing. *J. Nanosci. Nanotechno.* 1(2009) 350-357.

[12] V. O. Rouslan, D. F. I. Jeremy, M. S. Andrew. Whole Serum BSA Antibody Screening using a Lable-free Biophotonic Nanoparticle Array. *Anal. Biochem.* 385(2009) 234-241.

[13] V. Morosini, T. Bastogne, C. Frochote, et al.. Quantum dot-folic acid conjugates as potential photosensitizers in photodynamic therapy of cancer. *Photochem. Photobiol. Sci.* 10(2011) 842-851.

[14] Gao X, Yang L, J.A. Petros, et al.. In vivo molecular and cellular imaging with quantum dots. *Curr Opin Biotechnol.* 16(2005) 63-72.

[15] Li XG, Li F, He ZC. Large off-resonant and intensity-dependent third-order optical nonlinearities of CdSeS quantum dots/polystyrene composite film. *Materials Letters.* 93(2013) 366-369.

[16] He ZC, Li F. Optical properties of third-order nonlinearities of CdSeS quantum dots/Polystryrene composite film. *Chninese Journal of Lasers.* 1007001(2013) 1-5.

[17] I. L. Mendintz, H. Tetsuouyeda, E. R. Goldman, et al.. Quantum dot bioconjugates for imaging, labelling and sensing. *Nat. Mater.* 4(2005) 435-446.

[18] C.A. Leatherdale, W. K. Woo, F.V. Mikulec, et al.. On the absorption cross section of CdSe nanocrystal quantum dots. *J. Phys Chem.B.* 106(2002) 7619-7622.

[19] Gao X, Cui Y, R.M. Levenson, et al.. In vivo cancer targeting and imaging with semiconductor quantum dots, *Nat Biotechnol.* 22(2004) 969-976.

[20] Shao L J, Gao Y F, Yan F. Semiconductor quantum dots for biomedicial applications. *Sensors.* 11(2001) 11736-11751.

[21] A. C. Samia, X. Chen, C. Burda, Semiconductor Quantum Dots for Photodynamic Therapy. *J. Am. Chem. Soc.* 125(2003) 15736.

[22] L. X. Shi, B. Hernandez, M. Selke, Singlet Oxygen Generation from Water-Soluble Quantum Dot−Organic Dye Nanocomposites. *J. Am. Chem. Soc.* 128(2006) 6278.

[23] J. M. Hsieh, M. L. Ho, P. W. Wu, et al.. Iridium-complex modified CdSe/ZnS quantum dots: a conceptual design for bifunctionality toward imaging and photosensitization. *Chem. Commun.* 6(2006) 615–617.

[24] R. L. Sutherland. *Handbook of Nonlinear Optics.* Science Applications International Corporation, Dayton, Ohio, 1996.

[25] W. R. Algar, M. Massey, U. J. Krull,. The Application of Quantum Dots, Gold Nanoparticles and Molecular Switches to Optical Nucleic-Acid Diagnostics. *Trends Anal. Chem.* 28(2009) 292–306.
[26] T. Franzl, T. A. Klar, S. Schietinger, et al.. Exciton Recycling in Graded Gap Nanocrystal Structures. *Nano Lett.* 4(2004) 1599–1603.
[27] I. L. Medintz, A. R. Clapp, H. Mattoussi, et al.. Self-Assembled Nanoscale Biosensors Based on Quantum Dot FRET Donors. *Nat. Mater.* 2(2003) 630–638.
[28] A. L. Rogach, T. A. Klar, J. M. Lupton, et al.. Energy Transfer with Semiconductor Nanocrystals. *J. Mater. Chem.* 19(2009) 1208–1221.
[29] A. L. Rogach, Ed. Semiconductor Nanocrystal Quantum Dots. *Springer: New York,* 2008.
[30] A. U. Acuña; A. G. Francisco; M, Purificación et al. Structure and Formation of the Fluorescent Compound of Lignum nephriticum. *Organic Letters*. 11 (2009) 3020–3023.
[31] B. Valeur, S. Berberan, M. R. N. A Brief History of Fluorescence and Phosphorescence before the Emergence of Quantum Theory. *Journal of Chemical Education*. 88 (2001) 731.
[32] Edward Daniel Clarke. Account of a newly discovered variety of green fluor spar, of very uncommon beauty, and with remarkable properties of colour and phosphorescence. *The Annals of Philosophy.* 14(1819) 34-36.
[33] Haüy, Traité de Minéralogie, *2nd ed. Paris, France: Bachelier and Huzard*, vol1 (1822).
[34] David Brewster. On the colours of natural bodies. *Transactions of the Royal Society of Edinburgh*. 12(1834) 538-545.
[35] Herschel, John. On a case of superficial colour presented by a homogeneous liquid internally colourless. *Philosophical Transactions of the Royal Society of London*. 135(1845) 143-145.
[36] G. G. Stokes, M.A., F.R.S. On the Change of Refrangibility of Light. *Philosophical Transactions of the Royal Society of London*. 142(1852) 463-562.
[37] Friedrich Goppelsröder. Ueber eine fluorescirende Substanz aus dem Kubahol. *Journal for Praktische Chemie,* 101(1867) 408-414.
[38] C. A. Parker. Photoluminescence of Solutions, *Elesevier, Amsterdam.*
[39] Wodd. *Physical Optics.*
[40] J. Frank, G. Hertz. Über Zusammenstösse zwischen Electronen und den Molekülen des Quecksilberdampfes und die Ionisierungsspannung desselben, *Verhandlungen der Deutschen Physikalischen*. 16(1914) 457.
[41] G. Cario, J. Frank. *Physik,*1922.

[42] Wawillow, Stand und Aufgaben der Sprachwissenschaft, 1924.

[43] E. Gaviola. On Time-Lags in Fluorescence and in the Kerr and Faraday Effects. *Phys. Rev.* 33(1929) 1023.

[44] Courtesy of Dr. Ignacy Gryczynski.

[45] J. Szudy, ed. 1998. Uniwersytet Mikolaja Kopernika, Torun, Poland.

[46] L. Stryer. Fluorescence energy transfer as a spectroscopic ruler. *Annu Rev Biochem.* 47(1978) 819-846.

[47] Li F, Lu PX, Long H, et al.. Nonlinear absoprption in CuPc-doped PMMA thin film in the femtosecond regime: Experimental and theoretical studies. *Optic Express.*16(2008) 14571-14581.

[48] Li F, Li XG. Theoretical investigation on nonlinear absorption of multilevel organic molecular system in ns,ps and fs regime. *Optics Communications.* 285(2012) 5217-5222.

[49] J. R. Lakowicz. *Principles of Fluorescence Spectroscopy,* 2nd ed. Kluwer Academic: New York, 1999.

[50] N. J. Turro, Modern Molecular Photochemistry; University Science Books: Mill Valley, CA, 1991.

[51] A. R. Clapp, I. L. Mendinz, Mauro, J. M. et al.. Fluorescence Resonance Energy Transfer Between Quantum Dot Donors and Dye-Labeled Protein Acceptors. *J. Am.Chem.Soc.* 126(2004) 301-310.

[52] Li F, He ZC, Li MY, et al.. Three-photon absorption of copper phthalocyanine solution by femtosecond Z-scan technique. *Materials Letters.* 111(2013) 81-84.

[53] Li F, He ZC. Study of the reverse saturate absorption in copper phthalocyanine solution. *Advanced Materials Research.* 815(2013) 165-170.

[54] He ZC, Li F, Li MY, et al.. Fluorescence resonance energy transfer between CdTe quantum dots and copper phthalocyanine. *Acta Phys. Sin.* 64(2015) 046802.

[55] Li F, He ZC, Li MY, et al.. Photoinduced energy transfer in a CdTe quantum dot-copper phthalocyanine system via two-photon excitation. *Materials Letters.* 132(2014) 263-266.

[56] R. F. Kubin, A. N. Fletcher. Fluorescenc quantum yields of some rhodamine dyes. *Journal of Luminescence.* 27(1983) 455-462.

[57] Kim HN, Lee MH, Kim HJ, et al.. A new trend in rhodamine-based chemosensors:application of spirolactam ring-opening to sensing ions. *Chemical Society Reviews.* 37(2008) 1465-1472.

[58] Yu W.W, Qu LH,Guo WZ, Peng XG. Experimental Determination of the Extinction Coefficient of CdTe, CdSe, and CdS Nanocrystals. *Chem. Mater.* 15(2003) 2854-2860.

[59] Li MY, Li F, He ZC, et al. Two-photon-excited fluorescence resonance energy transfer in an aqueous system of CdTe quantum dots and Rhodamine B. *Journal of Applied Physics.* 233106(2014) 1-6.

In: Advances in Applied Physics Research ISBN: 978-1-63483-226-7
Editor: Gabriel Waters

Chapter 3

BAND OFFSETS IN DIELECTRIC/$InGaZnO_4$ THIN FILM TRANSISTORS

David C. Hays[1], Zachary McCracken[1], Matthew Anderson[1], B. P. Gila[1], S. J. Pearton[1] and Hyun Cho[2]

[1]Department of Materials Science and Engineering, University of Florida, Gainesville, FL, US

[2]Department of Nanomechatronics Engineering, Pusan National University, Gyeongnam, Korea

ABSTRACT

In recent years, amorphous oxide semiconductors (AOS) have attracted much attention for use in flat panel displays and flexible thin film transistors (TFTs) due to their improved electrical and optical performance as compared to amorphous Si, which is the dominant technology used in TFTs today. Several AOS materials have been studied including In-Zn-O, In-Sn-O, Zn-Sn-O, Zn-In-Sn-O, and In-Ga-Zn-O (a-IGZO). Among all AOS materials, a-IGZO has been the most widely investigated due to maintaining a very high electron mobility in the amorphous state (10-50 $cm^2 \bullet V^{-1} \bullet sec^{-1}$), and the wide controllability of carrier concentration through oxygen partial pressure during deposition. Moreover, its ability to be deposited at room temperature allows for use of novel flexible substrates, which raises the possibility of making low-cost electronics on a very wide range of arbitrary surfaces including plastic and paper. Amorphous IGZO TFTs are highly transparent, and can be deposited uniformly over large areas, opening up potential applications as switches in the active-matrix and driver-integrated circuits of low cost flexible displays. However, a major remaining issue with IGZO is a better understanding of the appropriate choice of compatible gate dielectrics for devices fabricated using IGZO channels. There are three main criteria that a gate dielectric must possess to be considered acceptable. First, it must be thermodynamically stable with the semiconductor and not react during processing. Secondly, it must provide a high quality interface with low defect and trap density to ensure high carrier mobility. Finally, it must act as a barrier to both electrons and holes, which is among the most important physical parameters for a given heterojunction system. Because these discontinuities can form a barrier for carrier transport across the interface, the knowledge of these quantities is essential for calculating the transport properties of the interface, or the electrostatic potential in a heterojunction device. It is recognized that it is essential to understand the energy band structure at heterojunction interfaces. While there has been much work to date on IGZO devices with various gate dielectrics, there has been very little work on determining the band offsets on such devices. In fact, in two recent reviews of advances in AOS TFTs, it was found that limited gate dielectrics have been studied and more work is needed for next generation TFT devices. The aim of this work is to systematically investigate various high-k dielectrics on IGZO and determine whether there is a sufficient barrier height at the interface in order to minimize the leakage current. This work is important for choosing the proper gate dielectric in next generation of TFT devices.

INTRODUCTION

Since 2004, there has been great interest in using amorphous oxide semiconductors (AOS) as the active layer for TFT's. Specifically, the ternary oxide system InGaZnO, has shown much promise. IGZO is a wide band gap (~3.5eV), n-type semiconductor and Table 1 highlights that it is the only material to combine high optical transparency, high carrier mobility with high yield and low cost.

Single-crystalline IGZO (sc-IGZO) has a complex, layered structure with alternating stacks of InO_2 and GaO(ZnO) [1-28] as shown in Figure 1. Unlike binary ZnO, a-IGZO can have a uniform amorphous phase because multiple oxides (In_2O_3 and Ga_2O_3) are introduced to promote glass phase formation [29], and the amorphous phase has been shown to be stable up to ~500˚C [18]. While several methods have been reported for a-IGZO deposition, such as

PLD [18, 30], it is RF [19] or DC [31] sputtering that is the most attractive due to their high deposition rate and high degree of uniformity.

Table 1. Comparison of TFT materials

TFT properties	α–IGZO TFT	α–Si TFT	poly–Si TFT	Organic TFT
Field effect mobility (cm^2 /V·s)	3 - 35	0.5 - 1	30 - 300	~0.1
Process Temperature (°C)	RT	~350°C	~450°C	<150°C
Transparency (%)	>80	<20	<20	>80
Yield	high	high	medium	high
Manufacturing cost	Low	Low	High	Low
Long term TFT reliability	High	Low	High	Low in air
Substrate	Glass Plastics Paper	Glass	Quartz	Glass Plastics

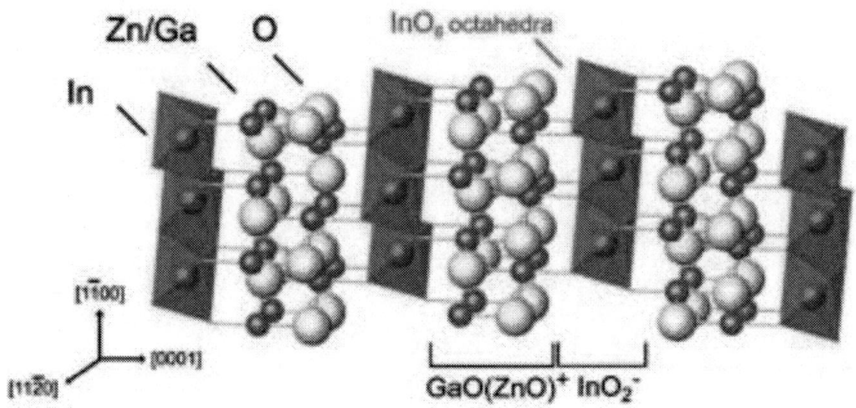

Figure 1. Structure of single crystal InGaZnO4 [adapted from ref. [27].

As illustrated in Figure 2 (top), in conventional, covalently bonded semiconductors (such as Si), electrons conduct through highly directional *sp3* hybrid bonding. When these semiconductors are in the amorphous state, the electron path is distorted by the resulting disorder and form deep localized states. This results in the electrons or holes traveling by hopping, which significantly reduces carrier mobility. On the other hand, in a-IGZO, the electrons are conducting through a metal ion's *s* orbital which is highly symmetrical. The high degree of symmetry allows the conducting path and carrier mobility to still be maintained even in the amorphous phase. (Figure 2 (bottom)) To ensure a high mobility in the amorphous phase, a sufficient *s* orbital overlap between metal ions is necessary. To satisfy this requirement, Hosono et al. [32-34] proposed a working hypothesis which predicts the metal ion should be heavy, post transition metal cations with a large principal quantum number ($n \geqq 5$). For example, the Hall mobility of the In-Ga-Zn-O system is primarily determined by the fraction of In_2O_3 content [32] in agreement with the Hosono theory since In^{3+} has the largest ionic radius ($n = 5$) among the metal cations.

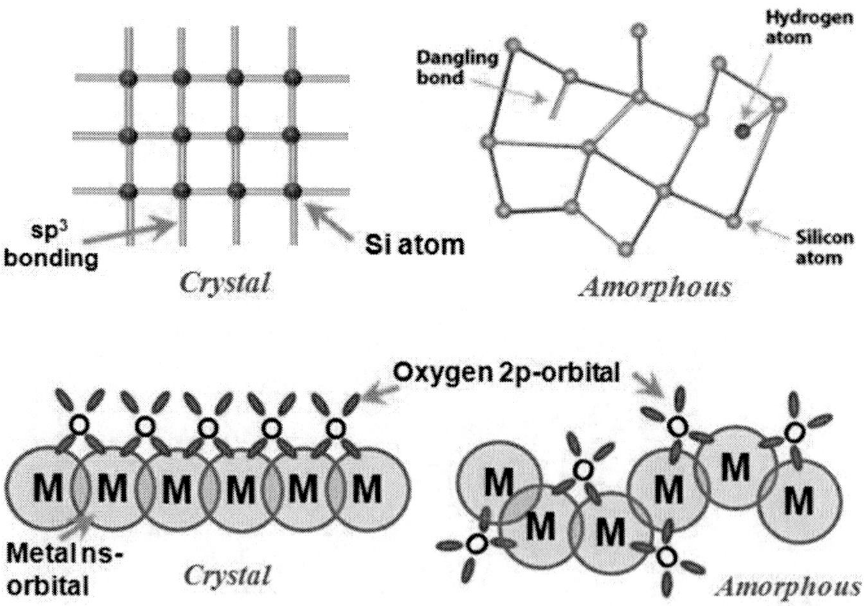

Figure 2. Schematic orbital drawing of electron conducting pathway (conduction band bottom) in (a) conventional covalent bond semiconductors (e.g., Si) and (b) ionic oxide semiconductors. (Adapted from ref. [18]).

For a TFT to have a low drain current in the off-state, and high on-to-off ratio, it is important to control the semiconductor carrier concentration to a very low level. In AOS films, carrier generation can occur via oxygen vacancy formation. The incorporation of Ga^{3+} helps by suppressing oxygen vacancies in a-IGZO thin-films since it has a stronger bond to oxygen than Zn or In ions [18]. A comparative study between a-IZO and a-IGZO has also shown that a-IGZO is able to have five orders of magnitude larger reduction in carrier concentration than a-IZO when two thin-films of these materials are deposited under the same condition, as shown in Figure 3.

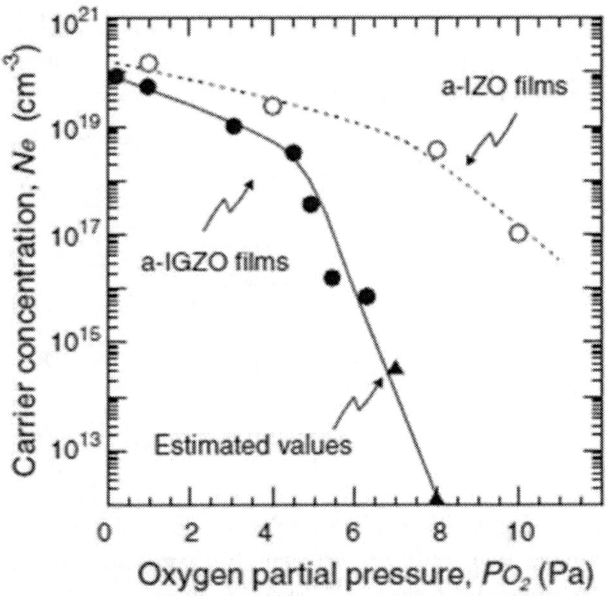

Figure 3. Dependence of carrier concentration on oxygen partial pressure for a-IGZO and a-IZO films (adapted from ref. [32]).

Finally, though a-Si:H is widely used in TFT backplanes today, it is highly absorbing in the visible spectrum with transmittance less than 30%. This has been a major drawback for utilizing such material in optoelectronics and there are also concerns with light induced instability [35]. On the other hand, a-IGZO has a wide band gap (~3eV) and is highly transparent in the visible spectrum with transmittance over 90%. A comparison of the absorption coefficients shows a three order of magnitude difference between a-Si and a-IGZO [36]. This property allows for higher resolution, transparent displays with an equivalent transmittance.

BAND OFFSET IMPORTANCE

There are three main criteria that a gate dielectric must possess to be considered acceptable. First, it must be thermodynamically stable with the semiconductor and not react during processing. Secondly, it must provide a high quality interface with low defect and trap density to ensure high carrier mobility. Finally, it must act as a barrier to both electrons and holes, which is among the most important physical parameters for a given heterojunction system [37-39]. Because these discontinuities can form a barrier for carrier transport across the interface, the knowledge of these quantities is essential for calculating the transport properties of the interface, or the electrostatic potential in a heterojunction device. Figure 4 shows the various types of band alignments: type I, type II staggered, type III broken gap.

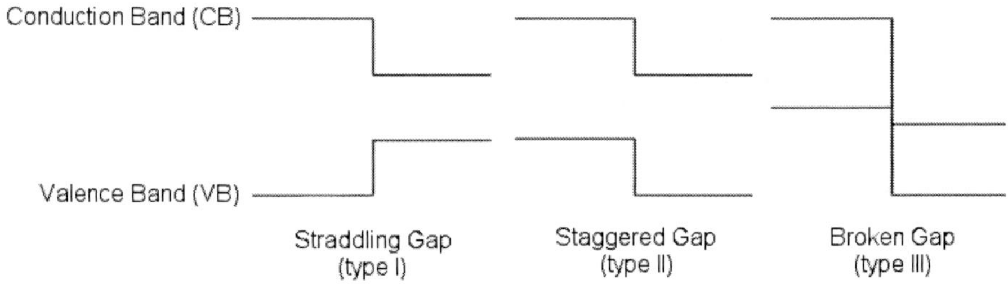

Figure 4. Types of band alignments.

The device concepts that can be implemented in a given heterojunction system will depend strongly on the type of band alignment, and device performance will depend on the exact values of the band offsets. Determining these alignments, unfortunately, is not possible using theoretical models since they cannot provide the accuracy needed for evaluating device concepts and must be determined experimentally.

The band diagram of a Metal-Oxide-Semiconductor (MOS) stack is a summary of the electronic structure of each material in the stack, and how it aligns with the electronic structure of the other layers. In a band diagram, the highest occupied state of the valence band is referred to as the valence band maximum (VBM), and is represented by a line. The lowest unoccupied state of the conduction band, which is also represented by a line, is the conduction band minimum (CBM). Figure 5 shows the band diagram for a typical MOS heterojunction. In this band diagram, the VBM and CBM of the oxide ($E_{V,OX}$, $E_{C,OX}$) and those of the semiconductor ($E_{V,S}$, $E_{C,S}$) are labeled.

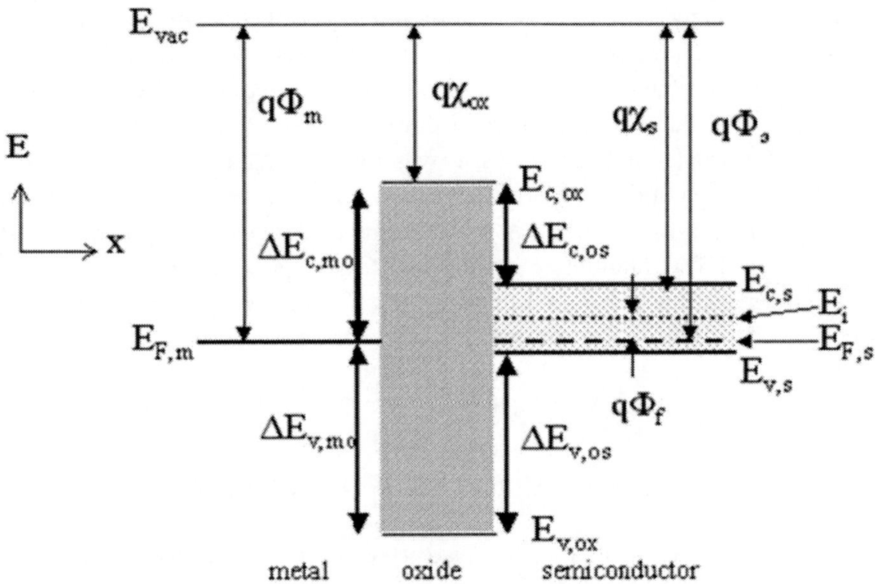

Figure 5. Band diagram of a Metal-Oxide-Semiconductor (MOS) stack.

If an electron at the bottom of the conduction band of the semiconductor were to move toward the oxide, it would, with a probability of 1 minus the tunneling probability, be reflected. This is because there are no electronic states in the oxide at this energy for it to occupy. Thus, the part of the oxide band gap above the Si CBM, which is equal to the difference between the oxide CBM and the semiconductor CBM, is the energy barrier to electrons in the semiconductor. This quantity is called the oxide-semiconductor conduction band offset (CBO), and is labeled $\Delta E_{C,OS}$ in Figure 5. Similarly, the difference between the oxide VBM and the Si VBM is the oxide-semiconductor valence band offset (VBO), which is labeled $\Delta E_{V,OS}$. This is the barrier that blocks the movement of holes from the semiconductor to the metal. In the same way, the movement of electrons from the metal to the semiconductor is blocked by the metal-oxide CBO ($\Delta E_{C,MO}$), which is the difference between the oxide CBM and the metal Fermi level ($E_{F,M}$). Finally, the metal-oxide VBO ($\Delta E_{C,MO}$), which is the difference between the oxide VBM and the metal Fermi level, blocks the flow of holes in the metal to the semiconductor. The probability for a charge carrier to tunnel through an energy barrier becomes non-trivial as the barrier is made very thin. The two factors on which the gate leakage current depends most strongly are the height of the energy barrier (VBO and CBO), and the thickness of the barrier.

It is recognized that it is essential to understand the energy band structure at heterojunction interfaces. While there has been much work to date on IGZO devices with various gate dielectrics, there has been very little work on determining the band offsets on such devices [40-51]. In fact, in two recent reviews of advances in AOS TFTs, it was found that limited gate dielectrics have been studied and more work is needed for next generation TFT devices [52, 53]. The aim of this work is to systematically investigate various high-k dielectrics on IGZO and determine whether there is a sufficient barrier height at the interface in order to minimize the leakage current. This work is important for choosing the proper gate dielectric in next generation of TFT devices.

DETERMINATION OF BAND OFFSETS

The most common method of measuring the VBO involves using X-ray photoelectron spectroscopy (XPS) to measure the core levels of a system. The basic method, shown in Figure 6, is to first measure the energy difference between a core level and the VBM for both a single layer dielectric and semiconductor of interest [54].

Heterojunction samples, consisting of a thin (~25A) layer of dielectric deposited on the semiconductor, are then prepared in which the separation between reference core levels in each material is measured. The separation between the reference core levels can be translated directly into a value for the valence band offset (VBO) using the previously measured single layer sample core-level to VBM energies:

$$\Delta E_V = \left(E^1_{core} - E^1_{VBM}\right) - \left(E^2_{core} - E^2_{VBM}\right) - \left(E^{\,1}_{\,core} - E^2_{core}\right)$$

The primary source of uncertainty in XPS measurements of band offsets arises from determining the position of the valence band maximum. The most used method produces a highly precise technique by linearly fitting the leading edge of the valence electron spectrum to the flat band energy, as shown in Figure 7.

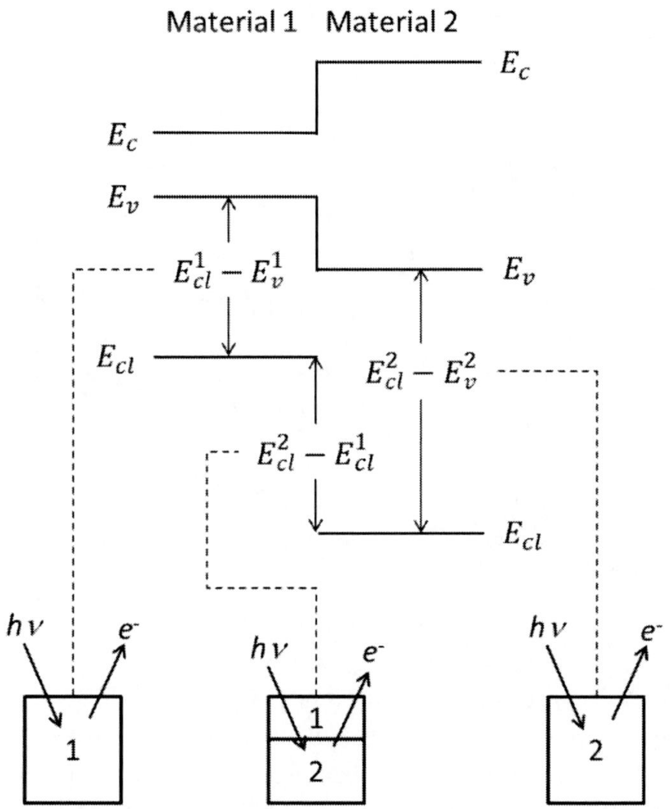

Figure 6. A schematic energy band diagram illustrating the basic principle of XPS band offset measurements.

To determine the conduction band offset, it is necessary to measure the band gap of each material. The most widely used method for measuring the optical band gap is UV/Vis spectroscopy. This technique measures the percent of light transmitted (%T) through a sample as a function of wavelength. To extract the band gap, the optical absorption coefficient, is calculated:

$$\alpha = -\frac{1}{d} ln(T)$$

where d is the thickness, and T, the ratio of transmitted (I) to the reference intensity (I_o). The band gap is related to the optical absorption coefficient in the following way:

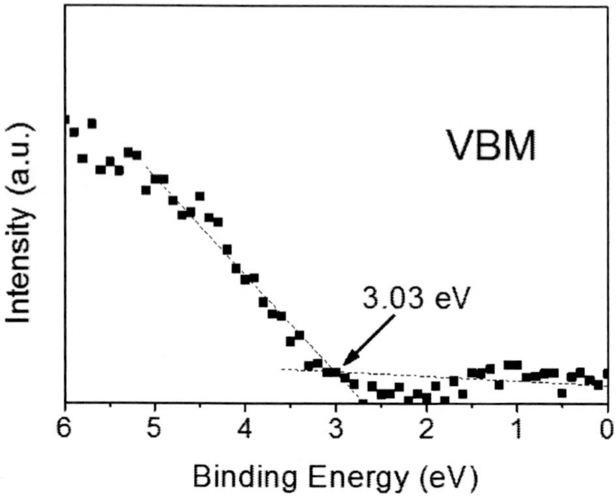

Figure 7. Linear fit of the valence band edge of HfO_2 on IGZO to determine the valence band maximum.

$$(\alpha hv)^2 = (hv - E_g)$$

where hv is the energy, α is the absorption coefficient from equation X, and E_g is the optical band gap. Thus to extract the band gap, one must plot $(\alpha hv)^2$ on the y-axis and hv on the x-axis. The band-gap energy (E_g) of the thin film can be obtained by utilizing the Tauc method [56] and extrapolating the linear portion of the curves relating $(\alpha hv)^2$ and hv to $(\alpha hv)^2 = 0$.

The UV-Vis spectral range is approximately 190 to 900 nm (~6.5eV to 1.4eV), as defined by the working range of typical commercial UV-Vis spectrophotometers. The short-wavelength limit for simple UV-Vis spectrometers is the absorption of ultraviolet wavelengths less than 180 nm by atmospheric gases. This lower wavelength limitation means that materials with a band gap greater than ~6.5eV cannot be measured using this technique. For band gaps greater than 6.5eV, XPS can also be used by measuring the onset of the loss feature on the high binding energy side of a core level [57]. To find the band-gap energy, a linear fit is made to the measured loss spectra curve near the approximate location of onset of inelastic losses. Next, by subtracting the background, the "zero" level is determined. The energy corresponding to the onset of inelastic losses is found by extrapolating the linear-fit line and calculating its intersection with the "zero" level. The bandgap energy is equal to the difference between the core-level peak energy and the onset of inelastic losses as shown in Figure 8. Thus, for oxides, the bandgap energy is calculated as, $E_g = E_{Loss} - E_{O1s}$. The precision of finding the band gap this way is limited, however, because the slope of the loss feature is not very different from that of the background of the XPS spectrum making background subtraction difficult.

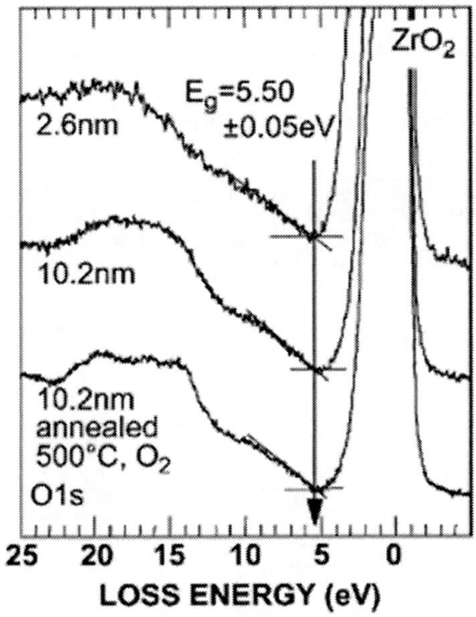

Figure 8. Measurement of the bandgap of Ta_2O_5 using onset of electron energy loss spectra.

Candidate Materials

High-k materials have been under investigation since the late 1990's in the silicon industry to identify those best suited to replace SiO_2 as the gate dielectric in MOSFETs. As stated previously, the primary criteria on which high-k materials are evaluated are dielectric constant, thermal and chemical stability in contact with the semiconductor, and the band offsets. The dielectric constant should be high as possible to be useful for scaling in the future, but often a material with a high dielectric constant has a smaller band gap and, consequently, smaller band offsets. Figure 9 shows the band gap versus dielectric constant of a number of high-k oxide candidate materials [58]. To act as an effective barrier, the offset at each band (CBO and VBO) should be greater than ~1eV to inhibit the injection of electrons and holes into the dielectric where they can cause threshold shifts and gate leakage.

Because IGZO has a band gap of ~3eV, suitable gate dielectrics should have a band gap of over 5eV to meet the minimum requirement for a 1eV band offset previously described. Figure 9 highlights some of these suitable materials in the shaded region. Based on this information. The properties of some of these candidate dielectrics are shown in Table 2.

Table 2. Properties of potential dielectrics for IGZO TFTs

Material	Band Gap (eV)	Dielectric Constant
SiO_2	8.9	3.8 – 5
MgO	7.8	8 – 10
CTMA	7.02	10
Al_2O_3	6.5	9 – 11
Y_2O_3	6.27	12 – 18
HfO_2	6.07	15 – 26
ZrO_2	5.8	25 – 30
YSZ	4.4	24 – 28

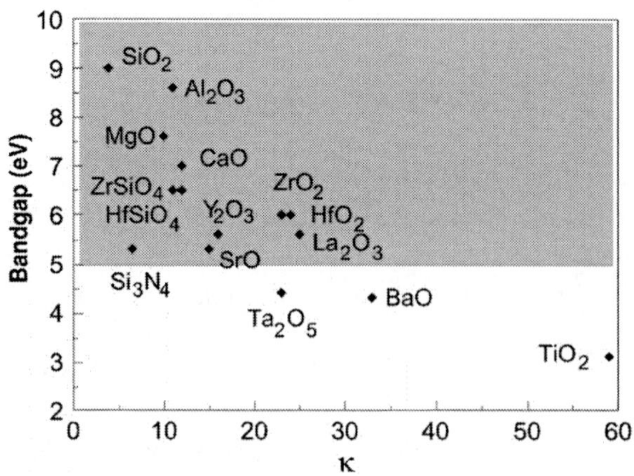

Figure 9. Dielectric constant (κ) versus bandgap for oxides.

DEPOSITION OF THIN FILMS

The primary method of depositing the thin films to be studied will be RF magnetron sputtering due to the availability of high quality, stoichiometric targets. In addition, as mentioned earlier, the deposition method of choice in industry is RF or DC sputtering due to its high deposition rate and high degree of uniformity, so this work will apply to tools and methods already in use.

Sputtering may be defined as the ejection of particles from a material when it is bombarded with high energy ions. When a large voltage is applied across a pair of electrodes in a low-pressure gas, the gas is ionized and a glow discharge or plasma is created [59]. The ions in the plasma are very energetic and being positively charged are attracted towards the cathode and cause the ejection or sputtering of cathode material. Argon is the gas most commonly used for sputtering, as it provides comparatively large ions for sputtering and since it is inert avoids any chemical reaction with the target. The sputtering yield depends mainly on the energy transfer from the incident particles to the cathode material, angle of incidence and also on whether the impinging particle is ionized or not [60]. One of the main advantages of the process is that sputtering is not strongly dependent on the cathode or the target temperature. This makes the process suitable for a very wide range of materials.

Planar magnetrons, which are the most common arrangement, were first introduced in the 1970s [61] to increase deposition rates and reduce substrate heating by confining the plasma to the target [62]. The cathode, commonly in the shape of a circular disc, is located above an arrangement of permanent magnets inside, which provides a field that confines the electrons above and close to the target. The cathode is connected to a power supply and the glow discharge is ignited. The region becomes densely populated with electrons from the ionized gas which under the influence of the Lorentz force from the magnets and the negative cathode potential travel along helical paths to the outer perimeter of the target where they are collected by a grounded anode preventing them from reaching the substrate. The configuration ensures a long path length for the electrons and as a result a high plasma density over the cathode with a correspondingly high number of positively charged ions, leading in turn to a high deposition rate.

DC sputtering cannot be used with insulators, as charge build-up at the cathode surface rapidly extinguishes the plasma and stops the sputtering process [63]. In such cases RF sputtering must be used, and a diagram of a basic RF sputter system is shown in Figure 13. In a RF discharge the electrons oscillate in the changing field thus gaining energy creating ions and more electrons by impact ionization. The insulating target is placed on the cathode, which has a metal backing plate. The plasma, the target and the metal plate act as capacitors with the potential at the surface of the target alternating between positive and negative. During the negative half cycle the positive ions bombard the surface of the target causing sputtering of the target. As the substrate holder, chamber walls, etc. are all grounded they effectively act as an anode of much greater area [64]. Since the capacitive reactance is inversely proportional to the area, a greater proportion of the potential will be dropped across the smaller cathode ensuring a higher field near the cathode and preventing ion bombardment of the substrate during the positive half cycle. The RF power supply operates at 13.56 MHz with peak-to-peak voltages above 1KV. To ensure maximum transfer of power to the plasma, it is generally necessary to include an impedance matching unit between the RF power supply and the cathode. Impedance matching generally consists of a number of variable reactive components, which must be adjusted to minimize the reflected power [65].

IGZO BAND OFFSET LITERATURE REVIEW

While many IGZO TFT devices have been studied in recent years [40-51], there has been very little work done on determining the band offsets of these materials. The band offsets are among the most important physical parameters for a given heterojunction system and critical for circuit designers to calculate the transport properties of the interface. A summary of the related works follows.

The SiO_2/IGZO system first was studied by Douglas et al., and was found to have strong carrier confinement due to the nested (Type 1) interface and large SiO_2 band gap, as shown in the top of figure 10 [66]. In addition, Tahir et al., performed a detailed study of the SiO_2/IGZO interface as a function of Ga content of the IGZO film [67, 68]. It was found that the IGZO band gap increased with Ga content, due to the higher band gap of Ga_2O_3, which caused the VBO to be shifted down. In comparison, the ratio of 1:1:1 (In:Ga:Zn) is the same material deposited by Douglas and can be directly compared as shown in Figure 10. The data are in close agreement in band gap values but have a ~0.5eV difference on band offsets which can be explained by either differences in charge compensation or different dielectric thickness, since the VBM displays a dependence on thickness [69].

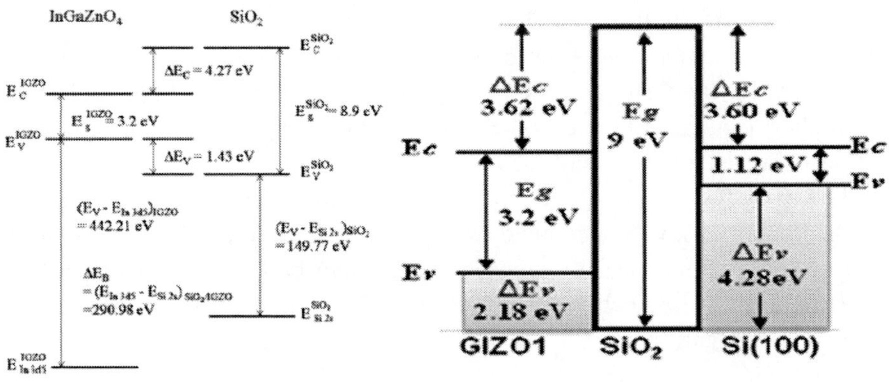

Figure 10. Energy band diagram of a thin SiO$_2$/IGZO heterojunction interface.

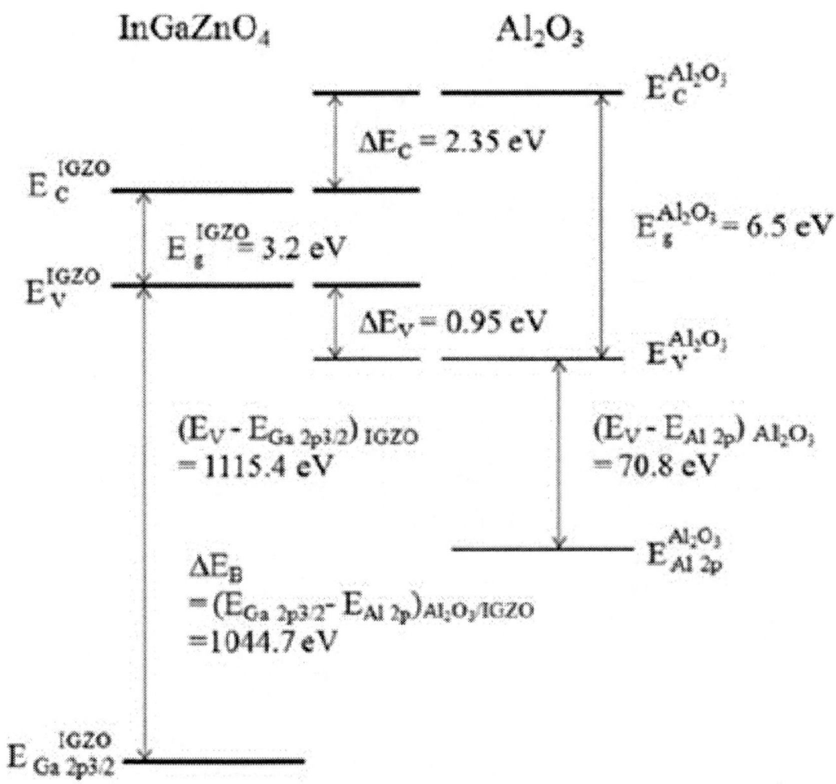

Figure 11. Energy band diagram of a thin Al$_2$O$_3$/IGZO heterojunction interface.

The band offsets for Al_2O_3/IGZO has been studied by Cho et al., and was determined to, again, have a nested (Type 1) interface with offsets of 2.35eV and 0.95eV for the CBO and VBO, respectively, as shown in Figure 11 [69]. The high offsets result in excellent carrier confinement, even though it is smaller than that achieved with SiO_2.

The ZrO_2/IGZO heterojunction was found by Yao et al., to have a nested interface but with no valence band offset as shown in Figure 12. [70]. With a CBO of 2.7eV, the material would be an excellent barrier to electrons but no barrier to holes. A low valence band offset could affect the stability of a device during negative gate voltage.

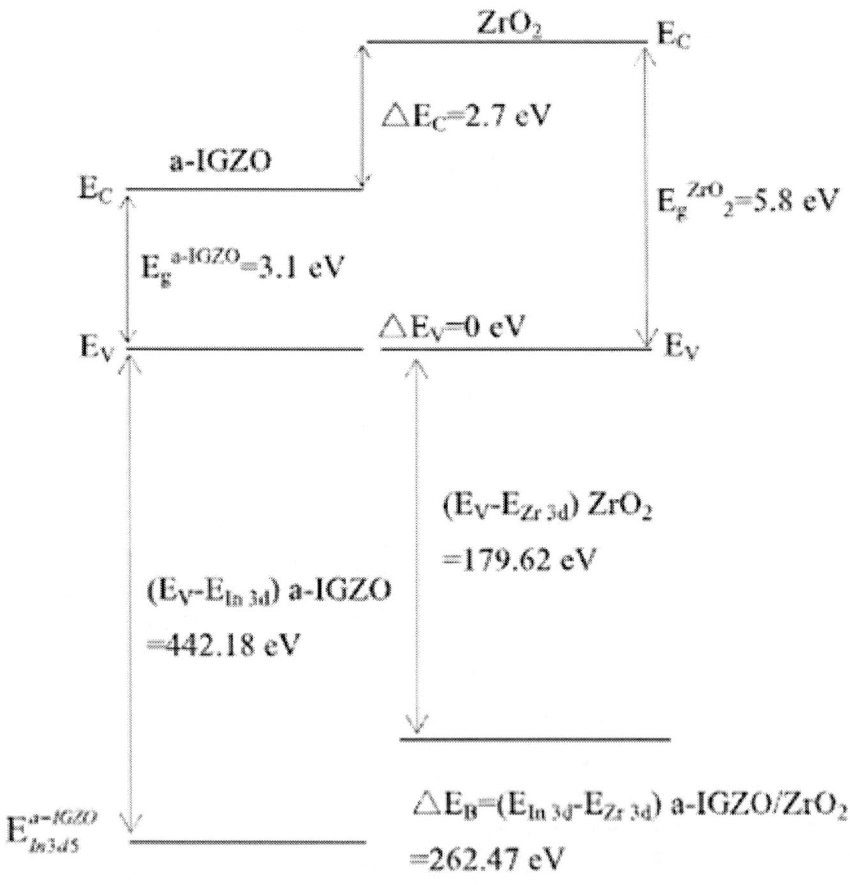

Figure 12. Energy band diagram of a thin ZrO_2/IGZO heterojunction interface.

A strong candidate material that has been well characterized and implemented in silicon IC manufacturing is HfO_2, due to its large dielectric constant and good thermal stability. This material was again studied by Cho et al., and was found to have sufficient carrier confinement with a CBO and VBO of 2.39eV and 0.48eV respectively [55].

The Y_2O_3-stabilized ZrO_2 (YSZ)/IGZO interface was studied by Kim et al., and was found to have a nearly perfectly symmetrical, nested band alignment, as shown in Figure 13 [71]. The lower band gap of YSZ produced smaller offsets of 0.63eV and 0.57eV for the CBO and VBO, respectively. Despite the decreased band offsets, the symmetrical alignment allows for good carrier confinement at low voltage driving conditions.

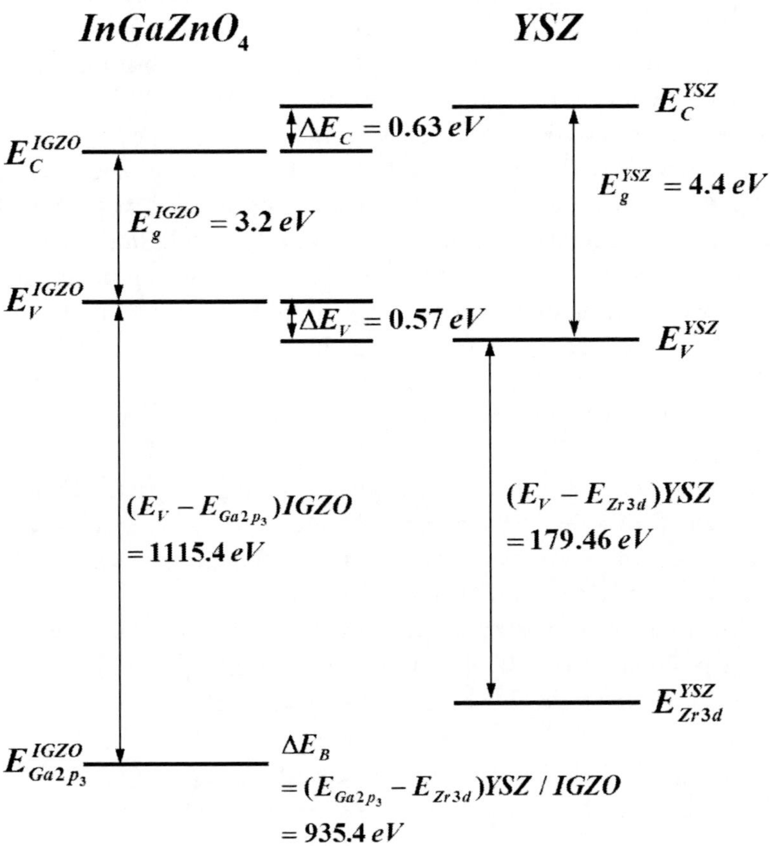

Figure 13. Energy band diagram of a thin YSZ/IGZO heterojunction interface.

CASE STUDY-HfO_2 BAND OFFSET

The HfO_2 was deposited by RF magnetron sputtering on InGaZnO/Si and quartz substrates at room temperature using a 3-in. diameter target of pure HfO_2. The RF power was 150W and the working pressure was 5 mTorr in a pure Ar ambient. The IGZO was deposited by sputtering on both Si and quartz by RF magnetron sputtering using a 3-in. diameter single target of $InGaZnO_4$. The RF power was 150 W, while the working pressure was constant at 5 mTorr in a pure Ar ambient.

To obtain the valence band offsets, XPS survey scans were performed to determine the chemical state of the IGZO and HfO2 films and identify peaks for high resolution analysis. Charge compensation was performed using an electron flood gun, due to the dielectric nature of the films. The charge compensation flood gun is often not sufficient at eliminating all surface charge, and additional corrections must be performed. Using the known position of the adventitious carbon (C-C) line in the C 1s spectra at 284.8 eV, charge correction was performed. A simple peak model involving a single C-C peak at 284.8eV was found to introduce enough error that I was unable to get similar valence band offsets to the paper from Cho. Additional peaks were added and the first was constrained to 1.5eV above the main peak and of equal FWHM. This higher binding energy peak is ascribed to alcohol (C-OH) and/or ester (C-O-C) functionality. A further high binding energy peak, attributed to O-C=O, was added with a position constraint of 3.7eV above the main peak. All peaks were constrained to a peak area ratio of 2:1:1 and are shown in Figure 14.

The valence band maximum (VBM) was determined by linearly fitting the leading edge of the valence band and the flat energy distribution, and finding the intersection of these two lines, as shown Figure 15. The VBM was measured to be 2.57eV for IGZO and 2.22eV for HfO_2.

High resolution XPS spectra were used to determine the selected core level peak positions for Hf $4f_5$, In $3d_5$, Ga $2p_3$, and Zn $2p_3$. These values were then inserted into the following equation to calculate ΔEv:

$$\Delta E_V = \left(E_{core}^{IGZO} - E_{VBM}^{IGZO}\right) - \left(E_{core}^{Hf02} - E_{VBM}^{Hf02}\right) - \left(E_{core}^{IGZO} - E_{core}^{Hf02}\right)$$

Again, it was not possible to reproduce the results previously obtained by Cho. An initial, unconstrained peak model was used for the Hf 4f double peaks which produced very good fits, but was found to not represent the actual surface. The Hf 4f7 and Hf 4f5 peaks are physically constrained to a peak area ratio of 4:3, with equal FWHM.

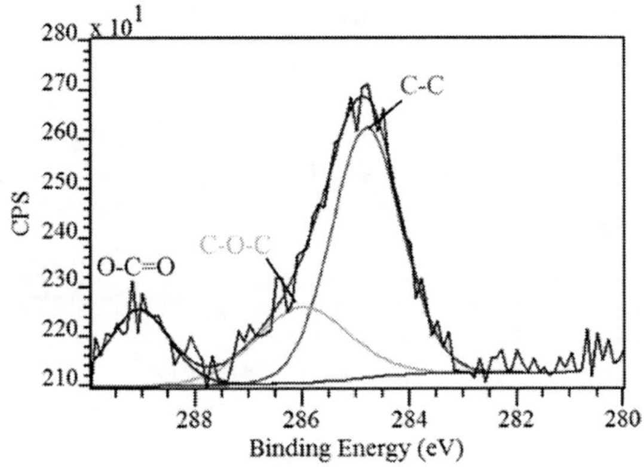

Figure 14. Optimized peak fit of C 1s spectra.

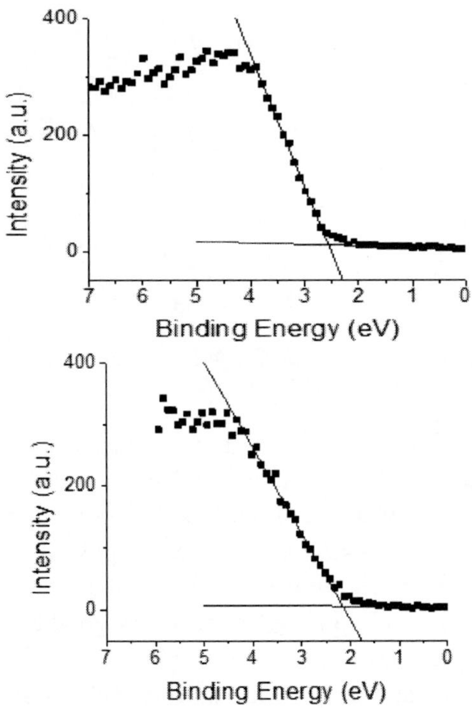

Figure 15. Linear extrapolation to determine valence band maximum of IGZO (top) and HfO_2 (bottom).

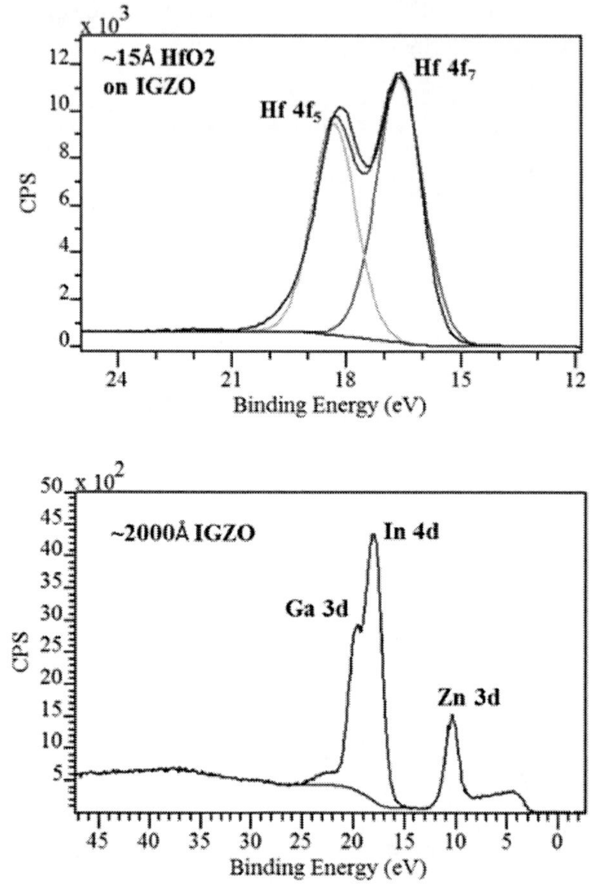

Figure 16. Constrained peaks for Hf 4f (top) showing poor fit to data because of peak contributions from the IGZO substrate (bottom).

With this constraint added, it was found that the Hf 4f5 peak could not reach the maximum as shown in Figure 16 (top), which implied another peak must be present. A detailed study of the IGZO XPS scans showed moderately high peaks for Ga 3d and In 4d that are directly lined up with the Hf 4f peak region, as shown in Figure 16 (bottom). The contribution from these peaks was added to the model and was able to produce a very good fit to the experimental data, as shown in Figure 17.

After peak optimization, we were able to determine the valence band offset to be ~0.38eV ± 0.03eV, which is within experimental error of the value of 0.47eV obtained by Cho.

The samples prepared on quartz substrates were used to determine the band gap of each material using UV/Vis spectroscopy. A Perkin-Elmer Lambda 800 dual beam, UV/Vis spectrometer was used over a wavelength range of 190-600nm (2-6.5eV). Each sample was simultaneously analyzed with a quartz reference substrate using the dual beam system to subtract out the substrate absorption effects. The optical band gap was determined using the method [56] described in section XX, and creating a Tauc plot then extrapolating the linear portion in the high energy region of the absorption edge.

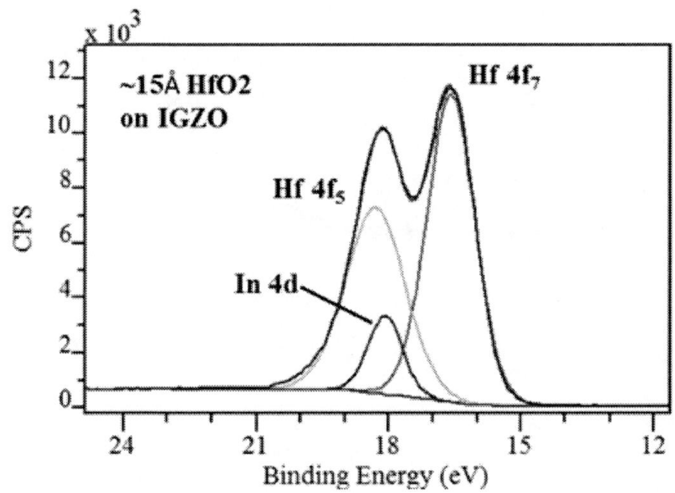

Figure 17. Optimized Hf 4f peaks with added contribution from IGZO substrate.

From the plots shown in Figure 18, a band gap of ~3.16eV and ~5.8eV was determined for IGZO and HfO2, respectively. The value for IGZO is in good agreement with that from Cho, Eg=3.2eV. However, there is almost a 0.2eV difference for the HfO2 band gap where he measured it to be 6.07eV. My results agree with previous reports of sputtered HfO_2 with the value ranging from 5.5-5.8eV [72, 73]. The difference between the data obtained by Cho could possibly be attributed to calibration issues or changes in stoichiometry. Using the measured band gap values, the conduction band offset (CBO) was determined to be 2.26eV using the following equation:

$$\Delta E_c = VBO + E_g^{IGZO} - E_g^{HfO2}$$

A summary of band offsets for different dielectrics on IGZO is shown in Figure 19. It is clearly possible to choose dielectrics which will provide good carrier confinement and low operating voltage operation of TFTs.

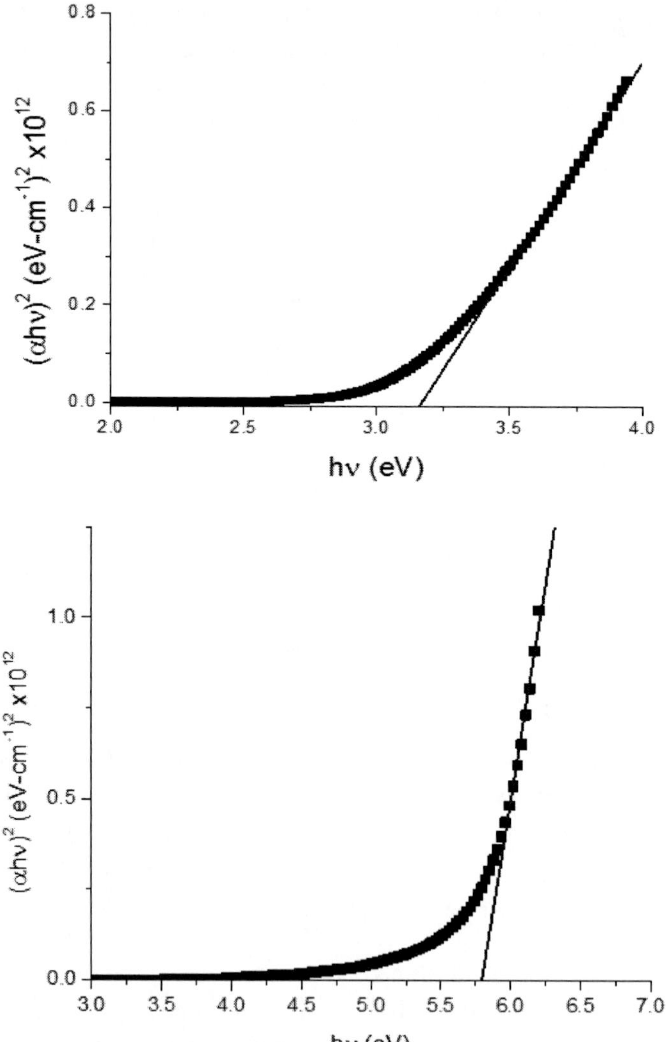

Figure 18. Tauc plot for IGZO (top) and HfO_2 (bottom) on quartz to determine the optical bad gap.

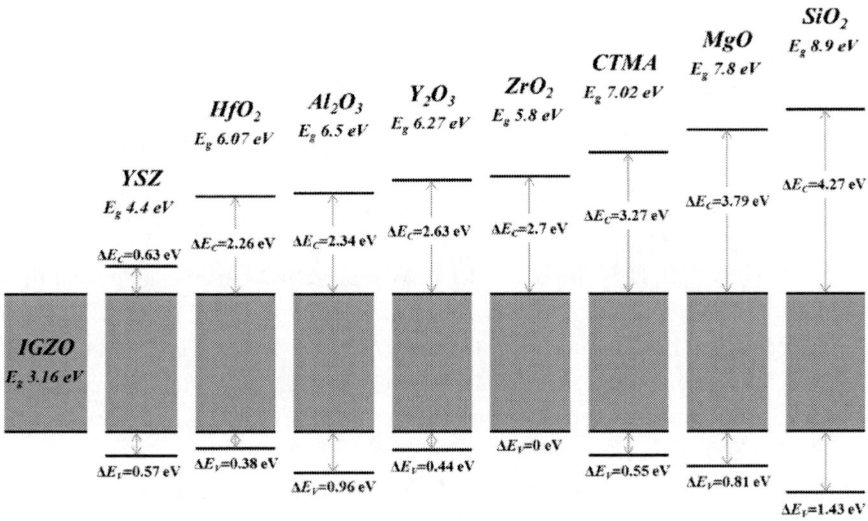

Figure 19. Summary of reported band offsets for dielectrics on IGZO.

SUMMARY AND CONCLUSION

In summary, the valence band offsets of many different dielectric/IGZO structures, including Y_2O_3/IGZO, HfO_2/IGZO, MgO/IGZO, Al_2O_3/IGZO, and SiO_2/IGZO heterojunctions were determined to be in the range of 0.44-1.43 eV from XPS measurements. Given the bandgap differences between each dielectric material and IGZO, conduction band offsets of 2.39-4.74 eV are more than sufficient to provide a strong carrier confinement in the IGZO channel. A key aspect to achieving good performance of these TFTs is the choice of gate dielectrics. The quality of dielectric/oxide interface plays an important role in TFT characteristics. A poor interface between the channel and gate-dielectric leads to large interface trap density and reduced carrier transport properties, which causes a degradation of device parameters such as leakage current, saturation mobility, and subthreshold gate-voltage swing. The work reported here is the beginning of a systematic study to establish the fundamental parameters of different dielectric/IGZO heterostructures.

ACKNOWLEDGMENTS

The work at UF was partially supported by NSF grant 1159682.

REFERENCES

[1] R.L. Hoffman, B.J. Norris, and J.F. Wager, "ZnO-based transparent thin-film transistors," *Applied Physics Letters*, 82, 733 (2003).

[2] S. Masuda, K. Kitamura, Y. Okumura, S. Miyatake, H. Tabata, and T. Kawai, "Transparent thin film transistors using ZnO as an active channel layer and their electrical properties," *Journal of Applied Physics*, 93, 1624 (2003).

[3] P.F. Carcia, R.S. McLean, M.H. Reilly, and G. Nunes, "Transparent ZnO thinfilm transistor fabricated by rf magnetron sputtering," *Applied Physics Letters*, 82, 1117 (2003).

[4] J. Nishii, F. M. Hossain, S. Takagi, T. Aita, K. Saikusa, Y. Ohmaki, I. Ohkubo, S. Kishimoto, A. Ohtomo, T. Fukumura, F. Matsukura, Y. Ohno, H. Koinuma, H. Ohno, and M. Kawasaki, "High mobility thin film transistors with transparent ZnO channels," *Japanese Journal of Applied Physics Letters Part 2*, 42, L347 (2003).

[5] E. Fortunato, A. Pimentel, L. Pereira, A. Goncalves, G. Lavareda, H. Aguas, I. Ferreira, C. N. Carvalho, and R. Martins, "High field-effect mobility zinc oxide thin film transistors produced at room temperature," *Journal of Non-Crystalline Solids*, 338-40, 806 (2004).

[6] E.M.C. Fortunato, P.M.C. Barquinha, A.C.M.B.G. Pimentel, A.M.F. Goncalves, A.J.S. Marques, L.M.N. Pereira, and R.F.P. Martins, "Fully transparent ZnO thin-film transistor produced at room temperature," *Advanced Materials*, 17, 590 (2005).

[7] N. L. Dehuff, E. S. Kettenring, D. Hong, H. Q. Chiang, J. F. Wager, R. L. Hoffman, C. H. Park, and D. A. Keszler, "Transparent thin-film transistors with zinc indium oxide channel layer," *Journal of Applied Physics*, 97, 064505 (2005).

[8] B. Yaglioglu, H. Y. Yeom, R. Beresford, and D. C. Paine, "High-mobility amorphous In_2O_3-10 wt%ZnO thin film transistors," *Applied Physics Letters*, 89, 062103 (2006).

[9] E. Fortunato, P. Barquinha, A. Pimentel, L. Pereira, G. Goncalves, and R. Martins, "Amorphous IZO TTFTs with saturation mobilities

exceeding 100 cm(2)/Vs," *Physica Status Solidi-Rapid Research Letters*, 1, R34 (2007).
[10] J. I. Song, J. S. Park, H. Kim, Y. W. Heo, J. H. Lee, J. J. Kim, G. M. Kim, and B. D. Choi, "Transparent amorphous indium zinc oxide thin-film transistors fabricated at room temperature," *Applied Physics Letters*, 90, 022106(2007).
[11] N. Itagaki, T. Iwasaki, H. Kumomi, T. Den, K. Nomura, T. Kamiya, and H. Hosono, "Zn-ln-O based thin-film transistors: Compositional dependence," *Physica Status Solidi a-Applications and Materials Science*, 205, 1915 (2008).
[12] T. Miyasako, M. Senoo, and E. Tokumitsu, "Ferroelectric-gate thin-film transistors using indium-tin-oxide channel with large charge controllability," *Applied Physics Letters*, 86, 162902 (2005).
[13] H. Q. Chiang, J. F. Wager, R. L. Hoffman, J. Jeong, and D. A. Keszler, "High mobility transparent thin-film transistors with amorphous zinc tin oxide channel layer," *Applied Physics Letters*, 86, 013503 (2005).
[14] P. Gorrn, P. Holzer, T. Riedl, W. Kowalsky, J. Wang, T. Weimann, P. Hinze, and S. Kipp, "Stability of transparent zinc tin oxide transistors under bias stress," *Applied Physics Letters*, 90, 063502 (2007).
[15] M. K. Jayaraj, K. J. Saji, K. Nomura, T. Kamiya, and H. Hosono, "Optical and electrical properties of amorphous zinc tin oxide thin films examined for thin film transistor application," *Journal of Vacuum Science & Technology B*, 26, 495 (2008).
[16] M. S. Grover, P. A. Hersh, H. Q. Chiang, E. S. Kettenring, J. F. Wager, and D. A. Keszler, "Thin-film transistors with transparent amorphous zinc indium tin oxide channel layer," *Journal of Physics D-Applied Physics*, 40, 1335 (2007).
[17] K. J. Saji, M. K. Jayaraj, K. Nomura, T. Kamiya, and H. Hosono, "Optical and carrier transport properties of cosputtered Zn-In-Sn-O films and their applications to TFTs," *Journal of the Electrochemical Society*, 155, H390 (2008).
[18] K. Nomura, H. Ohta, A. Takagi, T. Kamiya, M. Hirano, and H. Hosono, "Room-temperature fabrication of transparent flexible thin-film transistors using amorphous oxide semiconductors," *Nature*, 432, 488 (2004).
[19] H. Yabuta, M. Sano, K. Abe, T. Aiba, T. Den, H. Kumomi, K. Nomura, T. Kamiya, and H. Hosono, "High-mobility thin-film transistor with amorphous $InGaZnO_4$ channel fabricated by room temperature rf-magnetron sputtering," *Applied Physics Letters*, 89, 112123 (2006).

[20] T. Iwasaki, N. Itagaki, T. Den, H. Kumomi, K. Nomura, T. Kamiya, and H. Hosono, "Combinatorial approach to thin-film transistors using multicomponent semiconductor channels: An application to amorphous oxide semiconductors in In-Ga-Zn-O system," *Applied Physics Letters*, 90, 242114 (2007).

[21] K. Nomura, T. Kamiya, H. Ohta, M. Hirano, and H. Hosono, "Defect passivation and homogenization of amorphous oxide thin-film transistor by wet O_2 annealing," *Applied Physics Letters*, 93, 192107 (2008).

[22] S.-Y. Sung, J. H. Choi, U. B. Han, K. C. Lee, J.-H. Lee, J.-J. Kim, W. Lim, S. J. Pearton, D. P. Norton, and Y.W. Heo, "Effects of ambient atmosphere on the transfer characteristics and gate-bias stress stability of amorphous indium-gallium-zinc oxide thin-film transistors," *Applied Physics Letters*, 96, 102107 (2010).

[23] W. Lim, E.A. Douglas, D. P. Norton, S. J. Pearton, F. Ren, Y.W. Heo, S.Y. Son, and J.H. Yuh, "Improvement in bias stability of amorphous-$InGaZnO_4$ thin film transistors with SiOx passivation layers," *Journal of Vacuum Science Technology B,* 28, 116 (2010).

[24] W. Lim, J. H. Jang, S. H. Kim, D. P. Norton, V. Craciun, S. J. Pearton, F. Ren, and H. Shen, "High performance indium gallium zinc oxide thin film transistors fabricated on polyethylene terephthalate substrates," *Applied Physics Letters*, 93, 082102 (2008).

[25] J.K. Jeong, H.W. Yang, J. H. Jeong, Y.G. Mo, H.D. Kim, "Origin of threshold voltage instability in indium-gallium-zinc oxide thin film transistors," *Applied Physics Letters,* 93, 123508 (2008).

[26] Y.W. Heo, K. Cho, S. Sun, S. Kim, J. Lee, J. Kim, D.P. Norton and S.J. Pearton, "Effects of channel dimensions on performance of a-$InGaZnO_4$ thin-film transistors," *Journal of Vacuum Science Technology B*, 29, 021203 (2011).

[27] K. Nomura, T. Kamiya, H. Ohta, T. Uruga, M. Hirano, and H. Hosono, "Local coordination structure and electronic structure of the large electron mobility amorphous oxide semiconductor in-Ga-Zn-O: experiment and ab initio calculations," *Physical Review B*, 75, 35212 (2007)

[28] M. Orita, H. Tanji, M. Mizuno, H. Adachi, and I. Tanaka, "Mechanism of electrical conductivity of transparent InGaZnO4," *Physical Review B*, 61, 1811 (2000).

[29] M. Orita, H. Ohta, M. Hirano, S. Narushima, and H. Hosono, "Amorphous transparent conductive oxide $InGaO_3(ZnO)m$ ($m \leq 4$): a Zn4s conductor," *Philosophical Magazine B*, 81, 501 (2001).

[30] A. Takagi, K. Nomura, H. Ohta, H. Yanagi, T. Kamiya, M. Hirano, and H. Hosono, "Carrier transport and electronic structure in amorphous oxide semiconductor, a-InGaZnO4," *Thin Solid Films*, vol. 486, pp. 38-41, 2005.

[31] J. K. Jeong, J. H. Jeong, J. H. Choi, J. S. Im, S. H. Kim, H. W. Yang, K. N. Kang, K. S. Kim, T. K. Ahn, H.-J. Chung, M. Kim, B. S. Gu, J.-S. Park, Y.-G. Mo, H. D. Kim, and H. K. Chung, "12.1-inch WXGA AMOLED display driven by indium-gallium-zinc oxide TFTs array," *SID International Symposium Digest Tecnical Papers*, 1 (2008).

[32] H. Hosono, "Ionic amorphous oxide semiconductors: Material design, carrier transport, and device application," *Journal of Non-Crystalline Solids*, 352, 851 (2006).

[33] H. Hosono, N. Kikuchi, N. Ueda, and H. Kawazoe, "Working hypothesis to explore novel wide band gap electrically conducting amorphous oxides and examples," *Journal of Non-Crystalline Solids*, 198-200, 165 (1996).

[34] H. Hosono, M. Yasukawa, and H. Kawazoe, "Novel oxide amorphous semicon-ductors: Transparent conducting amorphous oxides," *Journal of Non-Crystalline Solids*, 203, 334 (1996).

[35] M. J. Powell, B. C. Easton, and D. H. Nicholls, "Annealing and light induced changes in the field effect conductance of amorphous silicon," *Journal Applied Physics*, 53, 5068 (1982).

[36] T.C. Fung, C.S. Chuang, K. Nomura, H.P.D. Shieh, H. Hosono, J. Kanicki, "Photofield-effect in amorphous In-Ga-Zn-O (a-IGZO) thin-film transistors," *Journal of Information Display*, 9, Iss. 4 (2008).

[37] J. Robertson, C.W. Chen, "Schottky barrier heights of tantalum oxide, barium strontium titanate, lead titanate, and strontium bismuth tantalite," *Applied Physics Letters*, 74 1168 (1999).

[38] J. Robertson, "Band offsets of wide-band-gap oxides and implications for future electronic devices," *Journal of Vacuum Science Technology B*, 18, 1785 (2000).

[39] J. Robertson, "Electronic Structure and Band Offsets of High-Dielectric-Constant Gate Oxides," *MRS Bulletin*, 217 (2002).

[40] Z.W. Zheng, C.H. Cheng, and Y.C. Chen, "Low Operation Voltage InGaZnO Thin Film Transistors with $LaAlO_3$ Gate Dielectric Incorporation," *ECS Journal of Solid State Science and Technology*, 2, N179 (2013).

[41] H.Y. Huang; Y.C. Huang; J.Y. Su; N.C. Su; C.K. Chiang; C.H. Wu; S.J. Wang, "High performance and low driving voltage amorphous InGaZnO

thin-film transistors using high-κ HfSiO dielectrics," *Device Research Conference (DRC)*, 235 (2010).

[42] W. Lim, Y. L. Wang, J. Lee, D.P. Norton, F. Ren, and S.J. Pearton, "High Performance $InGaZnO_4$-based Thin Film Transistors Fabricated at Low Temperature," *ECS Transactions*, 16, 303 (2008).

[43] T.M. Pan, C.H. Chen, J.H. Liu, F.H. Chen, J.L. Her, and K. Koyama, "Comparison of High-κ Gd_2O_3 and $GdTiO_3$ α-InGaZnO Thin-Film Transistors," *IEEE Transactions On Electron Devices*, 61, 87 (2014).

[44] J.S. Lee, S. Chang, S.M. Koo, and S.Y. Lee, "High-Performance a-IGZO TFT With ZrO_2 Gate Dielectric Fabricated at Room Temperature," *IEEE Electron Device Letters*, 31, 225 (2010).

[45] C.J. Chiu, S.P. Chang, and S.J. Chang, "High-Performance a-IGZO Thin-Film Transistor Using Ta_2O_5 Gate Dielectric," *IEEE Electron Device Letters*, 31, 1245 (2010).

[46] H.Q. Chiang, B.R. McFarlane, D. Hong, R.E. Presley, J.F. Wager, "Processing effects on the stability of amorphous indium gallium zinc oxide thin-film transistors," *Journal of Non-Crystalline Solids*, 354, 2826 (2008).

[47] Y.J. Cho, J.H. Shin, S.M. Bobade, Y.B. Kim, D.K. Choi, "Evaluation of Y_2O_3 gate insulators for a-IGZO thin film transistors," *Thin Solid Films*, 517, 4115 (2009).

[48] H. Kumomi, K. Nomura, T. Kamiya, H. Hosono, "Amorphous oxide channel TFTs," *Thin Solid Films*, 51, 1516 (2008).

[49] X. Ding, H. Zhang, J. Zhang, J. Li, W. Shi, X. Jiang, Z. Zhang, "IGZO thin film transistors with Al_2O_3 gate insulators fabricated at different temperatures," *Materials Science in Semiconductor Processing*, Available online 29 September 2013.

[50] H.H. Hsu, C.Y. Chang, C.H Cheng, S.H. Yu, C.Y. Su, C.Y Su, "Fully room-temperature IGZO thin film transistors adopting stacked gate dielectrics on flexible polycarbonate substrate," *Solid-State Electronics*, 89, 194 (2013).

[51] C.J. Chiu, S.P. Chang, C.Y. Lu, P.Y Su, S.J. Chang, "High-Performance a-IGZO Thin-Film Transistor with Organic Polymer Dielectric Layer," *AIP Conference Proceedings*, 1399, 929 (2011).

[52] J.S. Park, W.J. Maeng, H.S. Kim, J.S. Park, "Review of recent developments in amorphous oxide semiconductor thin-film transistor devices," *Thin Solid Films*, 520, 1679 (2012).

[53] E. Fortunato, P. Barquinha, and R. Martins, "Oxide Semiconductor Thin-Film Transistors: A Review of Recent Advances," *Advanced Materials*, 24, 2945 (2012).
[54] E.A. Kraut, R.W. Grant, J.R. Waldrop, and S.P. Kowalczyk. Phys. Rev. Lett., 44:1620, 1980.
[55] H. Cho, E.A. Douglas, B.P. Gila, V. Craciun, E.S. Lambers, F. Ren, and S.J. Pearton, "Band offsets in HfO_2/$InGaZnO_4$ heterojunctions," *Applied Physics Letters*, 100, 012105 (2012).
[56] H. Kusuma, Z. Hadi, I. Zuhairi, and K.S. Mohamad, "Optical Energy Gap of Ti:Al2O3 Single Crystals," *Journal of Applied Sciences*, 11.5, 888 (2011).
[57] S. Miyazaki, "Characterization of high-k gate dielectric/silicon interfaces," *Applied Surface Science*, 190, 66 (2002).
[58] J. Robertson, "Electronic Structure and Band Offsets of High-Dielectric-Constant Gate Oxides," *MRS Bulletin*, 27, 217(2002).
[59] K.L. Chopra, *Thin film phenomena*, McGraw-Hill Book Company, New York (1969).
[60] J.L. Vossen, W. Kern, "Thin film processes II," Academic Press, Boston (1991).
[61] J.S. Chapin, "Sputtering process and apparatus," *Research and Development Magazine*, 25, 37 (1974).
[62] A. Rockett, "Physical Vapor Deposition," *The Materials Science of Semiconductors*, Springer, pp. 505-572 (2008).
[63] B.S. Probyn, "Sputtering of insulators in a rf discharge," *Vacuum*, 18, 253 (1968).
[64] J.S. Logan, N.M Mazza, P.D. Davidse, "Electrical characterization of radiofrequency sputtering gas discharge," *Journal of Vacuum Science and Technology*, 6, 120 (1969).
[65] G.N Jackson, "RF Sputtering," *Thin Solid Films*, 5, 209 (1970).
[66] E. A. Douglas, A. Scheurmann, R. P. Davies, B. P. Gila, Hyun Cho, V. Craciun, E. S. Lambers, S. J. Pearton, and F. Ren, "Measurement of SiO_2 / $InZnGaO_4$ heterojunction band offsets by x-ray photoelectron spectroscopy," *Applied Physics Letters*, 98, 242110 (2011).
[67] D. Tahir, S.D.A. Ilyas, and H.J. Kang, "Band Alignment of Ultrathin GIZO/SIO_2/Si Heterostructure Determined by Electron Spectroscopy," *Makara, Sains*, 15, 193 (2011).
[68] Editors Gang He, Zhaoqi Sun "Observation of the interfacial dipole in high-k stacks," *High-k Gate Dielectrics for CMOS Technology*, John Wiley & Sons, p. 341 (2012).

[69] Hyun Cho, E.A. Douglas, A. Scheurmann, B.P. Gila,b, V. Craciun, E.S. Lambers, S.J. Pearton, and F. Ren, "Al_2O_3/$InGaZnO_4$ Heterojunction Band Offsets by X-Ray Photoelectron Spectroscopy," *Electrochemical and Solid-State Letters*, 14, H431 (2011).

[70] J. Yao, S. Zhang, and L. Gong, "Band offsets in ZrO_2/$InGaZnO_4$ heterojunction," *Applied Physics Letters*, 101, 093508 (2012).

[71] J.K. Kim, K-. W. Kim, E.A. Douglas, B.P. Gila, V. Craciun, E.S. Lambers, D.P. Norton, F. Ren, S.J. Pearton and H. Cho, "Band Offsets in YSZ/$InGaZnO_4$ Heterostructure System," *Journal of Nanoscience and Nanotechnology,* 14, 3925 (2014).

[72] F.L. Martınez, M. Toledano-Luque, J. J. Gandia, J. Carabe, W. Bohne, J. Rohrich, E. Strub and I. Martil, "Optical properties and structure of HfO_2 thin films grown by high pressure reactive sputtering," *Journal Of Physics D: Applied Physics*, 40, 5256 (2007).

[73] T. Tan, Z. Liu, H. Lu, W. Liu, F. Yan, W. Zhang, "Band structure and valence-band offset of HfO_2 thin film on Si substrate from photoemission spectroscopy," *Applied Physics A*, 97, 475 (2009).

INDEX

D

E

F

G

H

I

J

K

L

M

N

O

P

Q

R

S

T

U

V

W

X

Y

Z